Springer Business Cases

Springer Business Cases is a book series featuring the latest case studies in all areas of business, management, and finance, from around the world. The well-curated case collections in each of the books represent insights and lessons that can be used both in the classroom as well as in professional contexts. The books also place a focus on regional and topical diversity as well as encouraging alternative viewpoints which bring the knowledge forward. Both teaching cases as well as research cases are welcome.

Enrique Carlos Bianchi • Jose Luis Vazquez Burguete • M. Mercedes Galan-Ladero • Ana Lanero Carrizo
Editors

Responsible Consumption and Sustainability

Case Studies from Corporate Social Responsibility, Social Marketing, and Behavioral Economics

Springer

Editors
Enrique Carlos Bianchi
Faculty of Economic Sciences
National University of Córdoba
Córdoba, Argentina

Jose Luis Vazquez Burguete
Faculty of Economics and Business Sciences
University of León
León, Spain

M. Mercedes Galan-Ladero
University of Extremadura / Universidad de Extremadura
Badajoz, Spain

Ana Lanero Carrizo
Faculty of Economics and Business Sciences
University of León
León, Spain

ISSN 2662-5431 ISSN 2662-544X (electronic)
Springer Business Cases
ISBN 978-3-031-30741-6 ISBN 978-3-031-30742-3 (eBook)
https://doi.org/10.1007/978-3-031-30742-3

This Springer imprint is published by the registered company Springer Nature Switzerland AG
The registered company address is: Gewerbestrasse 11, 6330 Cham, Switzerland

"To all those who, through sustainable development,
seek to improve people's quality of life and build a better world."

"Be the change you wish to see in the world."
(attributed to M. Gandhi)

"Humanity is called to recognize the need for changes of
lifestyle, production and consumption"
(Pope Francis)

"You cannot get through a single day without having an impact on the world around you. What you do makes a difference, and you have to decide what kind of difference you want to make."
(J. Goodall)

Preface

Achieving economic growth and sustainable development requires that we urgently reduce our ecological footprint by changing the way we produce and consume goods and resources, and definitively our lifestyles. The efficient management of our shared natural resources and the way we dispose of toxic waste and pollutants are important targets to achieve this goal. Encouraging industries, businesses, and consumers to recycle and reduce waste is equally important, as is supporting countries to move toward more sustainable patterns of consumption by 2030.

Moreover, a large share of the world population is still consuming far too little to meet even their basic needs. Halving the per capita of global food waste at the retailer and consumer levels is also important for creating more efficient production and supply chains. This can help with food security and shift us toward a more resource-efficient economy (UNDP, 2022).

Currently, the importance of responsible consumption and sustainable lifestyle is unquestionable. Thus, responsible consumption is defined as "*the consumption that has less negative impact or more positive impact on the environment, society, the self, and the other-beings*" (Ulusoy, 2016). It is an umbrella concept that highlights various types of consumption terms that represent specific practices, such as sustainable consumption—consumers consider the impact of their consumption on society, the environment, and the economy and use resources taking future generations into consideration; ethical consumption—consumers are driven by pure motivations, though not every consumption that has ethical content impacts the well-being of others; consumer citizenship—consumers prioritize and actively contribute to the maintenance of just and sustainable development by caring and acting responsibly on family, local, regional, national, and global levels; socially responsible consumption—consumers prioritize their contribution to social issues and make it a point to buy from companies that care about these issues; and green consumption—consumers privilege their contribution to environmental issues over other topics.

On the other hand, sustainability has been considered as the triple bottom line of economic profitability, respect for the environment, and social responsibility. Sustainable living means understanding how our lifestyle choices impact the world around us and finding ways for everyone to live better and lighter. Applying a "people lens" to sustainability is new and timely and opportunities are great.

Sustainable living and lifestyles for the first time appear in the Sustainable Development Goals—SDGs (SDG 4—Education and SDG 12—Responsible Production and Consumption) (UNEP, 2020).

This book focuses on contributing to *Responsible Consumption and Sustainability* from a corporate social responsibility, social marketing, and/or behavioral economics perspectives, including cases from different regions around the world:

- By private, public organizations and hybrid organizations (B corporations).
- In countries such as Argentina, Chile, China, Mexico, Portugal, Spain, or in several countries.
- On topics such as:
  - Eco fashion design
  - Energy efficiency and renewable energy promotion
  - Environmental protection (client-customized use of natural resources, greenhouse gas reduction, packaging waste reduction and packaging reuse, and water management)
  - Regulation of the consumption (e.g., plastic materials)
  - Responsible production and consumption
  - Socially responsible consumption and sustainability
  - Sustainable brands and products (goods or services)
  - Sustainable business models
  - Sustainable development
  - Tourism sustainability promotion
  - Fair trade
- From different economic sectors, such as fashion, cosmetics, tourism, food industries (milk and wine), financial services, pharmaceutical industry, mining, petrol and energy, recycling industry, social economy, and higher education.
- Form different approaches, such as corporate social responsibility, (corporate) social marketing, circular economy, green marketing, and/or sustainability.

The objective of this book is to offer a useful book with high-quality international cases for students, researchers, marketers, and/or professionals. It includes the actions carried out by companies to minimize the social and environmental impact of the products (goods and services) they launch and the education campaigns that promote behavioral changes and new sustainable lifestyles.

Córdoba, Argentina — Enrique Carlos Bianchi
León, Spain — Jose Luis Vazquez Burguete
Badajoz, Spain — M. Mercedes Galan-Ladero
León, Spain — Ana Lanero Carrizo

Contents

Theoretical Background: Responsible Consumption and Sustainability—Corporate Social Responsibility, Social Marketing, and Behavioral Economic Approaches

Enrique Bianchi, Jose Luis Vazquez Burguete, M. Mercedes Galan-Ladero, and Ana Lanero Carrizo

1 Responsible Consumption and Sustainability

Ana Lanero-Carrizo and Jose Luis Vazquez Burguete

1.1 The Role of Responsible Consumption and Production in Sustainable Development

The evolution of society and economic activity, particularly during the twentieth century, has created conditions that threaten the environment at the planet level, as well as the prosperity and well-being of society (Belz et al., 2013). The world scenario is changing as a result of current growth trends. The concentration of economic power, the denationalization of economies, the forms of production and consumption, market systems, socioeconomic imbalances, environmental impacts, and the effects of the presence of human beings on the planet constitute an unprecedented legacy.

E. Bianchi (✉)
Universidad Nacional de Córdoba, Córdoba, Argentina
e-mail: enrique.carlos.bianchi@unc.edu.ar

J. L. Vazquez Burguete · A. Lanero Carrizo
Faculty of Economics and Business Sciences, University of León, León, Spain
e-mail: jose-luis.vazquez@unileon.es; ana.lanero@unileon.es

M. M. Galan-Ladero
University of Extremadura/Universidad de Extremadura, Badajoz, Spain
e-mail: mgalan@unex.es

E. C. Bianchi et al. (eds.), *Responsible Consumption and Sustainability*, Springer Business Cases, https://doi.org/10.1007/978-3-031-30742-3_1

Although significant advances have been made in science and technology, the daily ecological footprint is inexorably limiting the fate of future generations, so that the current trajectory of human development cannot be sustained and change is necessary.

In the year 2000, the beginning of a new century and the new millennium was presented as a unique opportunity to reflect on the state of the world and the prospects for the future of humanity. More than 20 years later, and especially as a result of the socio-sanitary and economic crisis caused by the COVID-19 pandemic, hardly a day goes by without a reference to "sustainability" in the media. However, the concept of *sustainable development* is not new or exclusive to recent years but the result of a process of progress and consolidation (Warde, 2018; Warde et al., 2018). The World Commission on Environment and Development (WCED), often referred to as the Brundtland Commission, provided in 1987 the most enduring and memorable definition of the term: "meeting the needs of the present without compromising the ability of future generations to meet their needs" (p. 24). Sustainable development is, thus, an alternative approach to the progress of humanity that tries to combine economic prosperity with social justice and the protection and quality of the environment (Belz et al., 2013).

This definition has opened a debate in recent decades about the impact of the production and marketing of goods and services on the environment, health, employment, education, and well-being. Thus, it seems clear that not only governments and public authorities are co-responsible for sustainable development but also companies and citizens; the effort must be joint and continuous (The United Nations, 2019). Specifically, companies must produce sustainable products, and citizens must be willing to consume them, so that finding new solutions that offer sustainable consumption and production modalities is in the interest of stakeholders throughout the value chain.

This reflection entrusts all types of organizations with the task of determining the interests of their target audiences, providing their satisfaction more effectively and efficiently than their competitors, but from a perspective of consumers as members of society, so that their needs can only be covered if the products (goods or services) offered to them do not have any harmful effect on the different areas in which they develop their lives (Kotler et al., 2011). Therefore, only those companies that intend to achieve their economic goals by controlling their harmful effects on society and the environment will be able to survive in the future (Daub & Ergenzinger, 2005).

In line with this approach, *sustainable marketing* is a type of marketing that is based on building lasting relationships with customers but in a particular frame of reference based on promoting sustainable development or considering sustainability issues. Thus, it is based on the analysis of the needs and wishes of the customers to develop sustainable solutions that provide superior value to them, as well as to communicate and distribute it effectively to the selected target groups, integrating throughout the entire process social and ecological aspects (Belz et al., 2013). In the words of Altinbasak-Farina and Burnaz (2019), the marketing activity of organizations must be governed by the four "S": customer satisfaction, product safety, social acceptability, and sustainability.

Organizations can also use their innovative power to design solutions that can inspire and motivate people to lead more sustainable lifestyles, reducing the effects and increasing well-being (The United Nations, 2022a). Thus, there are two main ways in which individuals can contribute to sustainable development: reduce waste and act thoughtfully and be informed when buying, opting for a sustainable option whenever possible (The United Nations, 2022a). All this has led to new profiles and types of responsible consumers and more sustainable lifestyles.

1.2 The Responsible Consumer and Sustainable Lifestyles

Involving companies in sustainable development is not an easy task, since it implies that the dimensions of environmental, social, and economic sustainability become part of the preferences of consumers in their purchasing decisions, and they are able to identify them in the products they purchase. In this sense, it is quite usual to think of consumption as an economic phenomenon that starts from what the individual wants and that directs the economy through the consolidation of collective behavior. But consumption is also a social and cultural process through which we express our identity and establish our place in society. It is also a physical process that utilizes resources of all kinds, so the collective consequences of consumption decisions and the way our needs are met are the main causes of climate change, which will have consequences for people, countries, and species from around the world (Belz et al., 2013).

Consumer concern about the impacts of their purchases is not, however, something new at the moment. As such, the concept of sustainable consumption formally arose in the second half of the twentieth century, parallel to that of sustainable development and, as a result of a cultural change, subsequent to modernization in Western countries (Llopis Goig, 2009). Reference began to be made, mainly, to a phenomenon of developed societies in which, once basic needs are covered, there is concern for issues related to the environment, civic and personal rights and liberties, and sociopolitical, intellectual, and aesthetic aspects.

In this context, during the 1980s, a debate started in Western Europe and North America on "green," "environmentally conscious," and "active" consumption, regarding a series of consumption patterns that were considered environmentally responsible. In industries with products as diverse as batteries, beverages, automobiles, cleaning products, detergents, and food, the environment has become a competitive factor. Thus, concepts of green marketing or environmental marketing arose, focused on the target group of "green" consumers who would be willing to pay higher prices for more ecological products (Peattie, 1992, 1995).

Unlike much of the ecological concern expressed during the 1970s, the prevailing concept of sustainable development in the twenty-first century is not based on a zero-growth agenda, but on ensuring that economic growth does not end up causing its own demise, since its consequences lead to serious social or environmental crises. As such, sustainable development provides a concept that business, politicians, and pressure groups can share and an ideological space in which a constructive dialogue could take place (Belz et al., 2013). In short, sustainable development poses the challenge of combining a dynamic economy with a society that offers opportunities for all (including future generations) that improves the productivity of resources but that slows down the growth of environmental degradation. This macroeconomic concept seeks to achieve worldwide social progress, stable levels of economic growth, employment, environmental protection, and natural resources (Morrós-Rivera & Vidal-Martínez, 2005).

From this renewed vision, a broad concept of *responsible consumption* arises, defined as "the consumption that has less negative impact or more positive impact on the environment, society, the self, and the other-beings" (Ulusoy, 2016, p. 285). The concept covers different types of consumers and consumption practices, such as the following (Ulusoy, 2016, p. 285):

- Sustainable consumption, referring to those consumers who "consider the impact of their consumption on society, the environment, and the economy, and who use resources taking future generations into consideration"
- Ethical consumption, defending that "consumers are driven by pure motivations (...) though not every consumption that has ethical content impacts the well-being of others"
- Consumer citizenship, referring to consumers who "prioritize and actively contribute to the maintenance of just and sustainable development by caring and acting responsibly on family, national and global levels"
- Socially responsible consumption, identifying consumers who "prioritize their contribution to social issues and make it a point to buy from companies that care about these issues" (p. 285)
- Green consumption, referring to consumers who "privilege their contribution to environmental issues over other issues"

In brief, the *responsible consumer* is one who expresses "concern about the effects generated by their purchasing activities and issues as, for example, the origin of the products, how they are obtained, or the actors directly or indirectly involved in the manufacture of goods or the provision of services. This is a movement influenced by values such as solidarity, social responsibility, respect for human rights, multiculturalism, and ecology, as well as rationally and directly linked to the education, ethics and maturity of individuals who claim their rights through social movements, as consumer platforms or organizations" (Vázquez & Lanero, 2021, p. 22).

In fact, sustainable consumption is being postulated as a fundamental priority within the current far-reaching policies around the world, as a way to combat the negative effects that the lifestyle of recent years is generating in terms of social inequality, loss of biodiversity, and climate change.

The United Nations Environmental Program (2018) recognizes this need and expresses it in clear and concise terms: "sustainable living means understanding how our lifestyle choices impact the world around us and finding ways for everyone to live better and lighter. Applying a 'people lens' to sustainability is new, timely and opportunities are great." Lifestyle initiatives can also help set the foundation for deeper changes required in policy and infrastructure contexts to design and deliver more sustainable living opportunities for people. This process not only involves the consumer but also depends on the responsibility of companies and entities in the cultural and socioeconomic context of consumption (Dueñas-Ocampo et al., 2014).

Fig. 1 Sustainable Development Goals. Source: The United Nations (2022a)

The importance of responsible consumption and sustainable lifestyles is included in the planning of the Sustainable Development Goals (SDGs) of the 2030 Agenda for Sustainable Development (The United Nations, 2022b).

2 SDG 12. Scope and Targets

Ana Lanero-Carrizo and Jose Luis Vazquez Burguete

2.1 What Are the SDGs?

The Sustainable Development Goals (SDGs) represent perhaps the most promising universal sustainable development blueprint agreed in recent times to end poverty, protect the planet, and improve the lives and prospects of the world's population (The United Nations, 2022b).The agreement consists of 17 objectives (Fig. 1), established from the work of the representatives of the 70 countries that participated in the Rio + 20 Conference in 2012. These SDGs were later established in the 2030 Agenda for Sustainable Development, contemplating an action plan to be achieved by the year 2030 with the purpose of calling for action by all countries—poor, rich, and middle-income—to promote prosperity while protecting the planet. It explicitly recognizes that ending poverty must go hand in hand with strategies that foster economic growth and address a range of social needs, including education, health, social protection, and employment opportunities, at the same time addressing climate change and environmental protection (The United Nations, 2015).

The 2030 Agenda takes as a reference the principles of sustainable development proposed in the Brundtland report (WCED, 1987), which recognizes the interdependence between the natural environment, human social well-being, and economic activity, as well as the need to establish and maintain a balanced dynamic between these three elements. The fundamental principles of sustainability are expressed implicitly and explicitly in the Brundtland definition (Belz et al., 2013) as follows:

- *Needs:* the approach shows both a concern for social justice and an anthropocentric concern, that is, it frames the maintenance of a healthy and sustainable environment as a necessary condition for human well-being.
- *Equity:* a call is made for economic development based on a fairer distribution of its costs and benefits between different countries, regions, races, and age groups and between the sexes.
- *Intergenerationality:* a long-term perspective is adopted, seeking a balance between the current needs and those of future generations, so that the needs of existing people are balanced with their ability to satisfy future generations of citizens, consumers, investors, and workers.
- *Global environmentalism:* it is recognized that the environment is a holistic, physical, dynamic, and vulnerable system with a limited capacity to provide resources to our system of production and consumption of resources and insufficient to absorb waste and pollution without impairing the quality of the environment and the utilities of all kinds that it provides.

From the 17 SDGs, 169 goals described in the Agenda emerge, aimed at solving the social, economic, and environmental problems that afflict the world, based on the collaboration of governments and public organizations, as well as the private sector, organizations, entities of civil society, and citizens. Although the SDGs are not legally binding, governments are expected to adopt them as their own and establish national frameworks for their achievement. According to the United Nations, countries have the primary responsibility for monitoring and reviewing progress toward meeting the goals.

2.2 SDG 12: Responsible Consumption and Production—Guarantee Sustainable Consumption and Production Patterns

The promotion of sustainable consumption is recognized as one of the development priorities, recognizing the power of the demand for more sustainable alternatives in the market as one of the main engines of production committed to social and environmental goals (The United Nations, 2022a).

SDG 12 is perhaps the objective most directly related to economic-business activity with regard to the production and marketing of products (goods and services). It urges "to guarantee sustainable production and consumption patterns"; that is, "to do more and better with less. It is also about decoupling economic growth from environmental degradation, increasing resource efficiency, and promoting sustainable lifestyles." Sustainable production and consumption also imply "the use of related products, which respond to basic needs and provide a better quality of life while minimizing the use of natural resources and toxic materials, as well as the emissions of waste and contaminants throughout the life cycle of the service or product so as not to jeopardize the needs of future generations" (The United Nations, 2022a).

Consumption and production, although the driving forces of the world economy, are totally dependent on the use of the natural environment and the planet's resources. In this sense, it can be said that the economic progress achieved in recent times has been accompanied by environmental degradation that has seriously threatened the very systems on which it depends. Just a few examples are enough (The United Nations, 2022a):

- If the world's population were to reach 9.6 billion by 2050, it would take the equivalent of nearly three planets to provide the natural resources needed to sustain today's lifestyles.
- The food sector represents around 30% of total energy consumption in the world and 22% of total greenhouse gas emissions.
- Although the most serious environmental impacts on food occur in the production and processing phase, households influence these impacts through their consumption habits and choices, which affects the environment through related energy consumption with food and waste generation.
- An estimated one-third of all food produced each year, equivalent to 1.3 billion tons worth around $1 billion, ends up rotting in consumer and retail bins or spoils due to poor shipping practices and the harvest.

Thus, it is noted that one of the greatest global challenges is to integrate environmental sustainability with economic growth and well-being and to decouple environmental degradation from economic growth. This segregation between resources and impact is complex but necessary to promote more efficient, sustainable patterns of production and consumption leading to a transition toward a greener and more socially inclusive world economy.

2.3 Main Objectives and Goals

Ensuring sustainable consumption and production practices necessarily implies respecting the biophysical limits of the planet, which means reducing current global consumption rates and adapting the biophysical capacity to produce, which benefits ecosystems. Sustainable consumption and production are therefore based on doing more and better with less. More specifically, Table 1 summarizes the specific targets of SDG 12.

Table 1 SDG 12 targets

Target 12.1: Implement the ten-year framework of programs on sustainable consumption and production patterns, all countries taking action, with developed countries taking the lead, taking into account the development and capabilities of developing countries
Target 12.2: By 2030, achieve the sustainable management and efficient use of natural resources
Target 12.3: By 2030, halve per capita global food waste at the retail and consumer levels and reduce food losses along production and supply chains, including post-harvest losses
Target 12.4: By 2020, achieve the environmentally sound management of chemicals and all wastes throughout their life cycle, in accordance with agreed international frameworks, and significantly reduce their release to air, water, and soil in order to minimize their adverse impacts on human health and the environment
Target 12.5: By 2030, substantially reduce waste generation through prevention, reduction, recycling, and reuse
Target 12.6: Encourage companies, especially large and transnational companies, to adopt sustainable practices and to integrate sustainability information into their reporting cycle
Target 12.7: Promote public procurement practices that are sustainable, in accordance with national policies and priorities
Target 12.8: By 2030, ensure that people everywhere have the relevant information and awareness for sustainable development and lifestyles in harmony with nature
Target 12.a: Support developing countries to strengthen their scientific and technological capacity to move toward more sustainable patterns of consumption and production
Target 12.b: Develop and implement tools to monitor sustainable development impacts for sustainable tourism that creates jobs and promotes local culture and products
Target 12.c: Rationalize inefficient fossil-fuel subsidies that encourage wasteful consumption by removing market distortions, in accordance with national circumstances, including by restructuring taxation and phasing out those harmful subsidies, where they exist, to reflect their environmental impacts, taking fully into account the specific needs and conditions of developing countries and minimizing the possible adverse impacts on their development in a manner that protects the poor and the affected communities

Source: The United Nations (2022a)

3 Different Approaches

3.1 Corporate Social Responsibility (CSR)

Enrique Bianchi

3.1.1 CSR Definition and History

The starting point of what is now referred to us as corporate social responsibility (CSR) and leads directly to modern debates about the nature and scope of the firm was the Berle-Dodd debate of 1932 (Berle, 1932). At that time, Merrick Dodd, in his paper entitled "For whom are corporate managers trustees," states that ". . .not only that business has responsibilities to the community but that our corporate managers who control business should voluntarily and without waiting for legal compulsion manage it in such a way as to fulfill those responsibilities" (Dodd, 1931).

During the 1950s, 20 years later, in "The Social Responsibilities of Businessmen," Bowen remarked that "the obligations of businessmen to pursue those policies, to make those decisions, or to follow those lines of action which are desirable in terms of the objectives and values of our society" (Bowen, 1953, p. 6).

In the 1960s, the concept of CSR underwent further elaboration and development, prompted by societal changes. It was the decade in which environmentalist movements were growing and there were increasing concerns about the Earth's ecosystem and pollution levels.

Close to the 1980s, Carroll (1979) implied that businesses had a commitment toward society and were obliged to engage in economic, legal, ethical, and discretionary (philanthropic) activities. In 1984, Freeman published *Strategic Management: A Stakeholder Approach* (Freeman, 1984).

Later, in 1991, Carroll refined his conception of CSR into a model which became known as the "Pyramid of Social Responsibility" which emphasizes the hierarchical dependencies between the four components (Carroll, 1991): economic, legal, ethical, and philanthropic responsibilities. In 1997, Elkington (1997) introduced the concept of the "triple bottom line" (3BL)—profit, people, and planet. It was prompted by an increasing awareness in, and concern for, "green" issues, sustainability, pollution, and the Earth's environment.

From that early beginning, CSR evolved in four chronological phases described in Table 2. In this four-stage emergence of CSR, firms may be positioned at different points on the phase trend line. Not all the companies pass through all four phases, although many of today's leading CSR enterprises manifest all of them simultaneously (Frederick, 2008).

The European Commission (2011, p. 6) considers that "respect for applicable legislation, and for collective agreements between social partners, is a prerequisite for meeting that responsibility. To fully meet their CSR, enterprises should have in place a process to integrate social, environmental, ethical, human rights and consumer concerns into their business operations and core strategy in close collaboration with their stakeholders."

Currently, there is still no consensus on a broad definition for CSR (Dahlsrud, 2008). On various occasions, the notion of CSR has been used as a synonym for business ethics, associated with corporate philanthropy, related to environmental policy or with corporate social performance or corporate citizenship (as we can see in Table 3).

3.1.2 Theories and Approaches

Garriga and Melé (2004) classify four types of CSR theories: (a) instrumental theories (focusing on maximizing shareholder value through social activities), (b) political theories (focusing on the political influence capacity of companies in society), (c) integrative theories (focusing on the integration of the social demands of the different stakeholders and the entire community), and (d) ethical theories (focusing on ethical behavior aimed at respecting people and their rights, the environment, and the common good of society).

Table 2 Evolving phases of CSR

Phases of corporate social responsibility		CSR drivers	CSR policy instruments	Authors and references
CSR 1 Early in the twentieth century but formally in the 1950s–1960s	Corporate social stewardship Corporate philanthropy—acts of charity Managers as public trustee-stewards Balancing social pressures	Executive conscience Company image/reputation	Philanthropic funding Public relations	Berle (1932) "For whom are corporate managers trustees" Bowen (1953) "Social responsibilities of the Businessman"
CSR 2 1960s–1970s	*Corporate social responsiveness* Social impact analysis Strategic priority for social response Organizational redesign and training for responsiveness Stakeholder mapping and implementation	Social unrest/protest Repeated corporate misbehavior Public policy/government regulation Stakeholder pressures Think tank policy papers	Stakeholder strategy Regulatory compliance Social audits Public affairs function Governance reform Political lobbying	Johnson (1971) "Business in contemporary society: framework and issues" Carroll (1979) "A three-dimensional conceptual model of corporate social performance" Friedman (1970) "The social responsibility of business is to increase its profits"
CSR 3 1980s–1990s	*Corporate/business ethics* Foster an ethical corporate culture Establish an ethical organizational climate Recognize common ethical principles	Religious/ethnic beliefs Technology-driven value changes Human rights pressures Code of ethics Ethics committee/officer/audits Ethics training Stakeholder negotiations	Mission/vision/values Statements CEO leadership ethics	Freeman (1984) "Strategic management: a stakeholder approach"
CSR 4 1990s–present	*Corporate/global citizenship* Stakeholder partnerships Integrate financial, social, and environmental performance	Global economic trade/investment High-tech communication networks Geopolitical shifts/competition Ecological	Intergovernmental compacts Global audit standards NGO dialogue Sustainability audits/reports	Carroll (1991) "The pyramid of corporate social responsibility: toward the moral management of organizational stakeholders" Elkington (1997) "Cannibals with forks: triple bottom line of

(continued)

Table 2 (continued)

Phases of corporate social responsibility		CSR drivers	CSR policy instruments	Authors and references
	Identify globalization impacts Sustainability of company and environment	awareness/ concern NGO pressures		twenty-first century business"

Source: Frederick (2008), and Crowther and Seifi (2021)

CSR theories are focused on four main aspects (Garriga & Melé, 2004): (a) meeting objectives that produce long-term profits, (b) using business power in a responsible way, (c) integrating social demands, and (d) contributing to a good society by doing what is ethically correct.

The strategic approach to CSR incorporates the perspectives arising out of institutional theory and stakeholder theory (Kao et al., 2016) and transaction cost theory (Acquier et al., 2017) (Table 4).

Finally, CSR strategies can be categorized into four key categories (Table 5): (a) dimensions, (b) benefits, (c) motivations, and (d) value creation and stakeholders.

3.1.3 Actions and CSR Initiatives

For Kotler and Lee (2005), the definition of CSR focuses on its commitment toward improving the well-being of the community and adding that there are strategic areas susceptible to CSR practices. The eight social initiatives (Table 6) exist from corporate, marketing, and communication and public relationship (Kotler et al., 2012; Vredenburg et al., 2020):

Corporate social initiatives:

- *Corporate philanthropy:* As voluntary donations of corporate resources to charitable causes: financial contributions, either in the form of direct grants or through vehicles like corporate foundations (Gautier & Pache, 2015)
- *Workforce volunteering:* Corporate programs that encourage employees to carry out community activities, contributing with their knowledge and/or skills (Haski-Leventhal et al., 2019)
- *Socially responsible business practices:* "Initiatives developed by companies—investments and business practices that support social causes—to improve the well-being of the community and protect the environment" (Kotler et al., 2012)

Marketing social initiatives:

- Cause promotion: "A corporation provides funds, in-kind contributions, or other corporate resources for promotions to increase awareness and concern about a social cause or to support fundraising, participation, or volunteer recruitment for a cause. The corporation may initiate and manage the promotion on its own... or it

Table 3 Main definitions of CSR

Bowen (1953, p. 6)	"What responsibilities to society may businessmen reasonably be expected to assume?" [CSR] refers to the obligations of businessmen to pursue those policies, to make those decisions, or to follow those lines of action, which are desirable in terms of the objectives and values of our society.
Davis (1960, p. 70)	[CSR] refers to the "businessmen's decisions and actions taken for reasons at least partially beyond the firm's direct economic or technical interest."
Frederick (1960, p. 60)	Social responsibility in the final analysis implies a public posture toward society's economic and human resources and a willingness to see that those resources are used for broad social ends and not simply for the narrowly circumscribed interests of private persons and firms."
Friedman (1962)	There is one and only one social responsibility of business—to use its resources and engage in activities designed to increase its profits so long as it stays within the rules of the game, which is to say engages in open and free competition without deception or fraud.
Johnson (1971, p. 50)	"A socially responsible firm is one whose managerial staff balances a multiplicity of interests. Instead of striving only for larger profits for its stockholder, a responsible enterprise also takes into account employees, suppliers, dealers, local communities, and the nation."
Carroll (1979, p. 500)	"The social responsibility of business encompasses the economic, legal, ethical and discretionary expectations that society has of organizations at a given point in time."
Jones (1980, pp 59–60)	"Corporate social responsibility is the notion that corporations have an obligation to constituent groups in society other than stockholders and beyond that prescribed by law and union contract. Two facets of this definition are critical. First, the obligation must be voluntarily adopted; behavior influenced by the coercive forces of law or union contract is not voluntary. Second, the obligation is a broad one, extending beyond the traditional duty to shareholders to other societal groups such as customers, employees, suppliers, and neighboring communities."
Wood (1991)	The basic idea of corporate social responsibility is that business and society are interwoven rather than distinct entities.
McWilliams and Siegel (2001, p. 1)	"CSR appears in situations where the firm goes beyond compliance and engages in actions that appear to further some social good, beyond the interests of the firm and that which is required by law."
Kotler and Lee (2005, p. 3)	"Corporate Social Responsibility is a commitment to improve community well-being through discretionary business practices and contributions of corporate resources."
Campbell (2007, p. 251)	"...Corporations act in socially responsible ways if they do two things. First, they must not knowingly do anything that could harm their stakeholders—notably, their investors, employees, customers, suppliers, or the local community within which they operate. Second, if corporations do cause harm to their stakeholders, they must then rectify it whenever the harm is discovered and brought to their attention."

Source: Based on Kakabadse et al. (2005)

Table 4 Theories and approaches of CSR

Theories and approaches	Description	Author(s)
Institutional theory	The companies are influenced by normative pressures, sometimes from external sources and sometimes from within the company itself, and the society acts to confer power on companies; thus, companies that fail to deploy their powers in manners that society deems responsible tend to lose them.	Zucker (1987) and Porter and Kramer (2006)
The stakeholder theory perspective	This approach considers that business can be connected with capitalism and ethics in order to create value for stakeholders. Three different aspects are considered: (a) how to understand and manage a company so as to generate wealth; (b) how to integrate ethics, responsibility, and sustainability from the perspective of capitalist economics; and (iii) what present and future managers should learn about successfully managing companies.	Freeman (1984) and Kao et al. (2016). Freeman et al. (2010)
Transaction cost theory	The transaction cost theory aims to minimize the costs for planning, executing, and monitoring the company's interactions with its agents and analyze the cooperation ongoing between companies and their noncommercial stakeholders (e.g., nongovernmental organizations, international associations, and local environmental associations) before concluding that transaction costs represent a factor explaining the difficulties encountered in implementing such CSR initiatives	Coase (1937, 1992)

Source: Nave and Ferreira (2019)

may be a major partner in an effort. . ., or it may be one of several sponsors" (Kotler et al., 2012, p. 22).

- *Cause-related marketing:* A corporation makes a donation (monetary or in-kind) based on the sales of its products (i.e., it ultimately depends on consumer behavior). Generally, this initiative is carried out over a given period of time, for a specific product and/or a specific NGO (Kotler & Lee, 2005, p. 23; Galan-Ladero et al., 2021, p. 4–5). The company usually partners with an NGO, in order to create a mutually beneficial ("win-win") relationship (i.e., on the one hand, the company will increase sales of its product; on the other hand, it also generates financial support for the charity and the cause it advocates).
- *Corporate social marketing:* "A corporation supports the development and/or implementation of a behavior change campaign intended to improve public health, safety, the environment, or community wellbeing. It is most distinguished by this behavior change focus, which differentiates it from cause promotions that focus on supporting awareness, fundraising, and volunteer recruitment for a cause. A corporation may develop and implement a behavior change campaign on its own. . . or more often it involves partners in public sector agencies and/or nonprofit organizations" (Kotler et al., 2012, p. 22).

Table 5 Main categories of CSR strategies

CSR		Key conclusions
Dimensions	Sustainability	Dimensions of sustainability are the following: – Economic: Is related to practices that contribute to the development of the region where the company operates. – Environmental: It means that organizations pay attention to not exceeding used resource capacities. – Social: Focus on initiatives that help resolve and prevent social problems and have been promoting social welfare in local and global communities.
	Ethics	Ethics represents a behavior adopted by organizations within which they balance their duties and business needs while also representing an adjustment between the company's culture and the values of society. Ethics constitutes one of the issues most valued by institutional and individual investors.
Benefits	Reputation	Reputation helps companies to: – Attract good members of staff; furthermore, workers achieve higher productivity and efficiency standards. – Increase the trust of investors.
	Performance	Organizations that implement CRS strategies: – Returns better financial performances than those maintaining traditional management strategies. – Influences the nonfinancial and financial performances of companies.
	Competitive advantage	Organizations applying socially responsible behaviors may develop sustained competitive advantages. CSR programs aid in building competitive advantages.
Motivation	Strategic motivations	A strategic motivation for the implementation of CSR can be linked with management issues.
	Altruistic motivations	The owner/managers' values are related with the integration of CSR in the organizational structure of the company.
Value creation and stakeholders	Internal	Managers of companies who engage in CSR activities can create value for their shareholders.
	External	CSR activities create value for the society and for the market.

Source: Nave and Ferreira (2019)

Communications and public relationship initiatives:

- *Advocacy advertising:* The main objective is to incite institutional change through shifting public opinion and behavior, defending themselves from criticism, mainly in harmful industries.
- *Brand activism:* Support a cause in order to encourage social political change and seek reputational and economic benefit via consumer appreciation of association with the cause.

Table 6 Corporate marketing and communications socially responsible initiatives

	Description	Form	Topic and degree of controversy
Corporate philanthropy	Making direct contributions to a charity or cause	Corporate funds and contributions	Non-controversial issue Social and community needs providing grants or donations to NGO Example: Providing grants for school field trips
Workforce volunteering	Supporting employees to volunteer in the community	Human resource programs	Non-controversial issue Social and community needs
Socially responsible business practices	Adapting and conducting discretionary business practices and investments that support	Business benchmark practices on sustainable development	Non-controversial issue Social and environmental issue related to the company chain value
Cause promotion	Supporting social causes through promotional sponsorships	Monetary (which also involves advertising)	Non-controversial charity, cause, or event. Example: Hosting book festivals for families to increase excitement about reading (Target Store)
Cause-related marketing	Linking monetary or in-kind donations to product sales or other consumer actions	Monetary (which also involves advertising)	Non-controversial charity, cause or event. Example: Making a contribution to water projects around the world with every purchase of Ethos Water at a Starbucks store
Corporate social marketing	Supporting behavior change campaigns	Advertising and social marketing campaign	Non-controversial issue affecting the corporation's immediate industry. Messaging only, no internal practice
Advocacy advertising	Incite institutional change through shifting public opinion and behavior	Public relationship and advertising	Issues involved with firms' operations, defending themselves from criticism, which mainly occurs in harmful industries such as oil and tobacco
Brand activism	Support a cause, raise awareness, and change behavior, reputational and economic benefit via consumer appreciation	Advertising and public relationship	Controversial socio-political cause, issue, charity, or event: Immigration, gender rights, LGBT, US gun reform, abortion and climate change. Alignment between messaging and practice

Source: Kotler et al. (2012) and Vredenburg et al. (2020)

3.2 Social Marketing

M. Mercedes Galan-Ladero

Social marketing,[1] also called *marketing of social causes*, or *marketing of ideas*, was first defined as "the design, implementation and control of programs calculated to influence the acceptance of social ideas and implies considerations of product planning, price, communication, distribution and marketing research" (Kotler & Zaltman, 1971, p. 5). Later, it was completed, considering that social marketing implies "the use of marketing techniques and principles to influence a specific public that voluntarily accepts, rejects, modifies or abandons behavior for the benefit of individuals, groups or society as a whole" (Kotler & Lee, 2005, p. 115).

Thus, its purpose is "to stimulate and facilitate the acceptance of ideas or social behaviors that are considered beneficial to society, or to stop or discourage those other ideas that are considered harmful" (Santesmases, 2007, p. 1036). Consequently, it focuses on social value creation, and its main aim is the search for social good (French & Gordon, 2020, p. 22), impacting significantly on people's quality of life (Alves et al., 2021, p. 3).

It has been used by all types of organizations (public and nonprofit organizations but also businesses—Galan-Ladero & Alves, 2019) to respond to social problems in different areas like health, environment, education, and also sustainability, among many others.

But some social problems are very complex and difficult to solve, requiring not only a change of behavior at the individual level but also changes in the environment and a transformation of social structures, as well as an involvement of authorities and decision-makers at the political level (Table 7).

On the other hand, approaches can also be classified according to their nature (Santesmases, 1999; Alves et al., 2021): legal, technological, economic, informational, and relational (Table 8). Normally, all these approaches are complementary.

On the other hand, it is also important to analyze the attitudes and behaviors of individuals toward the proposed ideas or social behaviors. The strategies to follow will depend on whether the attitudes are favorable or not and whether the socially desired behaviors are carried out or not (Table 9).

And regarding the planning of a social marketing program, Kotler and Lee (2008) proposed several steps (Table 10).

[1]For more information, see Alves et al. (2021). Theoretical Background: Social Marketing and Quality of Life. In Galan-Ladero, M.M., and Rivera, R.G. (Eds.) Applied Social Marketing and Quality of Life. Applying Quality of Life Research. Springer, Cham. https://doi.org/10.1007/978-3-030-83286-5_1.

Table 7 Combination of level approaches to solve social problems

Approaches	Definitions
Downstream approaches	Influencing the behavior of individuals directly: For example, *to reduce tobacco consumption: information campaigns about the health hazards of tobacco consumption* For example, *to reduce alcohol consumption: information campaigns about the health hazards of alcohol consumption* For example, *to recycle: information campaigns about the benefits of recycling and how to separate waste and which waste should go in which garbage container*
Midstream approaches	Acting upon those that can help behavior change (e.g., families, friends, coworkers, peers, etc.): For example, *pressuring them not to smoke, asking them not to smoke when they are all together* For example, *pressuring them not to drink alcohol, making them aware of the negative consequences of drinking alcoholic beverages, and/or setting an example* (e.g., *by not drinking alcoholic beverages at home or in front of the children)* For example, *teaching them to recycle, setting an example at home (separating the garbage and recycling the different types of waste)*
Upstream approaches	Influencing through social and environmental policies, changes in social structures, and involving decision-makers (e.g., politicians, authorities, experts, etc.): For example, *banning smoking in public places, restricting tobacco sales (only in authorized establishments), banning tobacco advertising, or raising the price of cigarettes* For example, *banning the sale of alcohol to minors, prohibiting the advertising of alcoholic beverages, carrying out awareness campaigns about the dangers of getting drunk*, etc. For example, *encouraging recycling, placing containers of different types close to homes, enacting laws on waste collection (with financial penalties if recycling is not done correctly or simply not collecting garbage that is not properly sorted), carrying out awareness campaigns*, etc.

Source: Own elaboration, adapted from Donovan and Henley (2010) and Basil et al. (2019)

3.3 Behavioral Economic Approach

Enrique Bianchi

It is possible to define behavioral economics (BE) as the study of economics that does not rely on the assumption of the rational, selfish economic person. Behavioral economics employs "theories and results from psychology, sociology, anthropology, neurology, and other disciplines, and makes use of empirical studies including experimental ones to demonstrate the inconsistency between the assumption of an *homo economicus* and actual economic decision making. Because there are important questions that cannot be addressed within a framework of traditional economics which relies on the supposition of a *rational selfish homo economicus*, this new approach attempts to leave this assumption behind" (Ogaki & Tanaka, 2017, p. 5).

BE research uses psychological experimentation to develop theories about human decision-making and has identified a range of biases. The field is trying to change the

Table 8 Complementary approaches to achieve the socially desired behavior

Approaches	Definitions
Legal approach	Through laws, regulations, rules, etc.: For example, *to reduce tobacco consumption: ban on smoking in public places, ban on tobacco sales to minors under 18 years of age, and ban on tobacco advertising.* For example, *to reduce alcohol consumption: ban on drinking in the street, ban on the sale of alcohol to minors,* etc. For example, *to recycle: laws on the collection of waste and disposal of hazardous materials*
Technological approach	Through innovations: For example, *medicines (nicotine gum, nicotine patches, pills,* etc.*)* For example, *medicines (to reduce alcohol addiction, to treat alcoholism), breathalyzers,* etc. For example, *smart containers (to warn when they are full, to compact the garbage inside, to recognize the type of waste, and even to identify who is depositing the waste)*
Economic approach	Through costs and prices (reducing the cost of performing the socially desired behavior, to encourage it, or increasing the cost of not performing it or of performing socially undesired behaviors, to discourage and curb them). For example, *increase the price of the pack of cigarettes, and maintain very high economic sanctions for establishments that sell tobacco to minors.* For example, *fines for drinking alcohol on the street, fines for selling alcohol to minors,* etc. For example, *tax discounts for correct recycling, tax payments based on garbage generated, fines for not recycling correctly,* etc.
Informational approach	Providing information (especially persuasive information): For example, *warning labels, images of different types of cancer caused by tobacco on cigarette packs, fear-appeal messages,* etc. For example, *with shocking images of the consequences of drinking alcohol (traffic accidents, drunken young people lying on the ground,* etc.*)* For example, *about products that can be created from recycled materials, about the impact that intensive, low-cost production* (e.g., *the fast fashion industry) and irresponsible consumption (following the throw-away culture) have on the environment, or the negative consequences of the excessive use of plastics on the earth, the oceans, animals, and even people's health.*
Relational approach	Based on interpersonal interactions: For example, *promoting tobacco-free lifestyles among teens (by the teens themselves, among their classmates, and friends) or by people talking about their experiences (how life changes when quitting smoking, how they have had to deal with tobacco-induced illnesses,* etc.*)* For example, *to make young people aware that it is not necessary to drink alcohol to have fun, through other young people or people who have had an accident as a result of driving after drinking alcohol,* etc. For example, *promoting recycling in schools (making children aware of the need to recycle, by teaching them to recycle at school, through their classmates and teachers).*

Source: Own elaboration, based on Alves et al. (2021) and Galan-Ladero and Alves (2023)

way economists think about people's perceptions of value and expressed preferences (Samson, 2014).

Table 9 Possible strategies to apply in social marketing, depending on the purpose sought

Strategy	Attitude	Behavior	Definition
Reinforcement	Positive	Done	To reinforce these ideas, attitudes, and/or behaviors: For example, with economic incentives (money) and noneconomic incentives (awards, rewards, social recognition, etc.) e.g., *receiving tax discounts, or gifts, for recycling*
Induction	Positive	Not done	To induce the accomplishment of a desirable behavior: For example, providing economic incentives, legal regulations, facilitating material and human resources to carry out the desired behavior e.g., *by installing containers near homes for recycling*
Rationalization	Negative	Done	To change the attitude: For example, giving more information, trying to persuade with campaigns, etc. e.g., *to inform about the advantages of separating waste so that it can be recycled (for those people who only recycle because if they do not, their garbage will not be picked up)*
Confrontation	Negative	Not done	To generate a change in attitude and behavior: For example, applying sanctions and coercive actions, using persuasive information, etc. e.g., *not to collect garbage that is not properly separated and, in addition, to fine its owners. Fining citizens who do not pick up their dogs' waste*

Source: Own elaboration, based on Santesmases (1999), Penelas-Leguía et al. (2012), Alves and Galan-Ladero (2019), and Alves et al. (2021)

3.3.1 Three Principles of Human Decision-Making

The three principles of human decision-making are (World Bank, 2015): thinking automatically, thinking socially, and thinking with mental models.

First Principle: Thinking Automatically

People cannot process all the information they receive on a daily basis to make decisions. For this reason, they recognize two types of decisions: those of a routine nature and those that are not (new, uncertain). These types of decisions are linked to two types of mental processes: (a) those that are fast, automatic, effortless, and associative and (b) those that are slower, reflective, and deliberative and therefore require more effort. Psychologists describe the two modes, metaphorically, as two distinct systems in the mind: System 1, the automatic system; and System 2, the deliberative system (World Bank, 2015). Individuals have two systems of thinking—the automatic system and the deliberative system. The automatic system influences nearly all our judgments and decisions (Table 11).

Table 10 Steps in the planning process for a successful social marketing program

Steps	Actions
1	*Define the problem, the purpose* (behavior to change), *and the focus* (who the target public is). Marketing research is key in this step.
2	*Conduct a situation analysis and summarize it in a tool like SWOT analysis*: strengths, weaknesses, opportunities, and threats.
3	*Select target audiences* (e.g., according to demographic and socioeconomic, psychographic, or behavioral criteria). Aspects to consider when selecting a target: who has the greatest need for change, who will be receptive to change, who is reachable, who is affordable, if the group size is sufficient/appropriate, if it fits with the organization, if it will be cost effective to reach. *Segmentation* is key, because people have different needs, values, interests, and motivations.
4	*Set marketing objectives and goals* (specifying desired behaviors and changes in knowledge, attitudes, and/or beliefs, e.g., reinforcing, inducing, rationalizing, or confronting them). *Objectives* should be clear, realistic, and measurable. *Goals* should be specific, measurable, achievable, relevant, and time sensitive.
5	*Identify factors influencing behavior adoption* (e.g., the barriers that prevent the audience of adopting the new behavior; the benefits they may realize when adopting the new behavior; the influencers of the desired behavior, who can help the persistence of the current behavior or help to change it; etc.). Barriers, benefits, influencers, facilitators, and "competitors" with your efforts.
6	*Craft a positioning statement* (according to the target audience's feelings and primary benefits sought when performing the desired behavior) to appeal the target.
7	*Develop marketing mix strategies (the 4Ps)*: product, price, place, and promotion.
8	*Outline a plan for monitoring and evaluation* (identifying which processes and outcomes will be measured, what methods will be used to measure them, when the measurements will take place, and the costs involved in carrying out the measurements).
9	*Establish budgets* (total costs of the program: including implementation costs, and control and evaluation costs), *and find funding sources* (to support the costs).
10	*Complete the plan for campaign implementation and management* (conceptualizing the outline of the task division: who will perform every task, how much it will cost, and when they should be carried out, including the partners' roles).

Source: Adapted from Kotler and Lee (2008), Alves and Galan-Ladero (2019), Basil et al. (2019), Alves et al. (2021), and Galan-Ladero and Alves (2023)

Second Principle: Thinking Socially

Individuals are social beings who are influenced by social preferences, social networks, social identities, and social norms: most people care about what those around them are doing and how they fit into their groups, and they imitate the behavior of others almost automatically.

What others think, expect, and do influences our preferences and decisions. Humans are inherently social. In making decisions, we are often affected by what others are thinking and doing and what they expect from us. Others can pull us toward certain frames and patterns of collective behavior.

Table 11 People have two systems of thinking

Automatic system	Deliberative system
Considers what automatically comes to mind (narrow frame)	Considers a broad set of relevant factors (wide frame)
Effortless	Effortful
Associative	Based on reasoning
Intuitive	Reflective

Sources: Kahneman (2003) and Evans (2008) cited in World Bank (2015)

Table 12 Examples of highly cost-effective behavioral interventions

Intervention	Description
Reminders	Weekly reminders through text messages or WhatsApps to patients to remind them to take the medicines prescribed by the doctors
Nonmonetary gifts	Small nonfinancial incentives and prizes—like lentils and metal dinner plates—were combined with a reliable immunization provider within the community
Public notices	Stickers or small signs indicating "How I drive" that allows you to report reckless drivers or drivers who drive drunk
Making products convenient	Point-of-collection water chlorination system, in combination with encouragement from community promoters, dramatically increased access to safe water
Inspirational messages	In vulnerable neighborhoods, the testimonies of people who had managed to escape extreme poverty and improve their economic and family situation are highly motivating and hopeful
Timing of cash transfers	Government school aid to families is paid if the infants attend school and have their vaccination card up-to-date

Source: World Bank (2015, p. 13)

Third Principle: Thinking with Mental Models

When people think, they generally do not draw on concepts that they have invented themselves. Instead, they use concepts, categories, identities, prototypes, stereotypes, causal narratives, and worldviews drawn from the individuals who do not respond to objective experience but to mental representations of experience. In constructing their mental representations, people use interpretive frames provided by mental models. People have access to multiple and often conflicting mental models. Using a different mental model can change what an individual perceives and how he or she interprets it. Table 12 presents examples of highly cost-effective behavioral interventions based on a more realistic understanding of human behavior.

3.3.2 Behavioral Bias and Behavioral Tools to Counter Biases

Humans have limited rationality and often do not act in our own best interest. Our daily decisions are guided by hundreds of behavioral biases. Many of these biases are exacerbated under stress and fatigue. The good news is that these errors are systematic and, therefore, can be corrected with simple interventions that reorient our decisions in the desired direction. Behavioral biases are systematic deviations from rational behavior.

Table 13 Behavioral bias

Behavioral bias	Description
Status quo bias or inertia	It refers to the human tendency to maintain the current state of affairs. This current state, or status quo, is taken as a reference point, and any deviation from it is perceived as a loss.
Cognitive overload	It is the amount of mental effort used at a given time. The overload occurs when the amount of offered information exceeds the individual's ability to process it. As our attention and memory are limited, we may be unable to process all available information.
Social norms	It refers to unwritten rules that govern behavior in any given society. It is important to differentiate between "descriptive norms" and "prescriptive norms." The former describes how individuals tend to behave (e.g., "most people arrive on time"). The latter defines what is considered acceptable or expected behavior, regardless of how individuals actually behave ("Please, arrive on time").
Overconfidence	It is the tendency to overestimate or exaggerate our own ability to successfully complete a task or to think we are better than the average.
Optimism bias	It leads us to underestimate the probability that negative events will happen to us and to overestimate the probability of positive events.
Loss aversion	It is the notion that the discomfort derived from a loss is perceived as bigger than the happiness derived from an equivalent gain.
Availability heuristic	It describes the tendency of individuals to estimate the probability of a future event based on how readily representative examples of such event come to mind.
Representativeness heuristic	It describes the tendency of individuals to estimate the probability of an uncertain event occurring based on how similar it is to other more certain events.
Hassle factors	They are seemingly small inconveniences, such as having to read a lot of information or needing to take an extra small step to complete an action, that can hinder or disrupt our decision-making process.
Time inconsistency or present bias	It is the tendency to be impatient when choosing between gains today or in the future but patient when choosing between gains in two different moments in the future.

Source: Martínez Villarreal et al. (2020)

The main behavioral biases and tools to counter biases are briefly described in Tables 13 and 14.

Finally, by designing strategies and nudges to correct biases and barriers that steer choices toward suboptimal outcomes, the researchers have helped people make wiser decisions in a number of areas, including education, saving, health, tax compliance, and labor market. Through "nudges," which are changes made to the choice architecture without forbidding any other options or significantly changing economic incentives (Thaler & Sunstein, 2008), behavioral economics can help generate systematic and predictable changes in the behavior of a group in a specific context (Table 15).

Table 14 Behavioral tools to encounter bias

Behavioral tools	Description
Framing	It is the tendency to reach different conclusions depending on how the information is presented. For example, decision options can be presented in a way that highlights either their positive or negative aspects, causing each option to be perceived as more (or less) attractive.
Salience	It refers to the tendency of individuals to focus on items or information that are more prominent and ignore those that are less so. Thus, it is important to make key aspects of messages stand out and to display them in an appropriate place and time.
Defaults	They automatically present actions that are triggered if nothing else is chosen by the decision-maker. Generally, they seek to overcome cognitive overload or present bias, supported by our tendency to maintain the status quo.
Commitment device	It is a choice we make in the present that restricts our future set of options to those that reflect our long-term objectives. This helps mitigate future impulsive behaviors. Leveraging our tendency to inertia, this tool can help address time inconsistency and cognitive overload.
Reminders	They can take many forms, such as an email, a text message, a letter, or a personal visit reminding individuals of some aspect of their decision-making. Reminders are designed to mitigate procrastination, oversight, and cognitive overload.
Micro-incentives	They are rewards or punishments designed to influence individuals' behaviors or decisions. They can be tangible, such as food or money, or intangible, such as public recognition. Unlike incentives, which are an integral part of public policy design, micro-incentives are small, low-cost, easy-to-apply cues that can complement the original policy design. This tool can be used to mitigate loss aversion and present bias.
Planning prompts	They are messages designed to encourage individuals to make a concrete action plan. Planning prompts motivate people to break down their goal (e.g., attending a doctor's appointment) into smaller, concrete tasks (e.g., leaving work early, finding a babysitter, postponing a weekly meeting, etc.) and to anticipate unforeseen events. These prompts often include space for the individual to write down relevant information such as the date, time, and place of their commitment.
Descriptive norms	They describe the behavior of a social group, regardless of whether it is positive or negative. Communicating these norms can help change behaviors. For example, one might think that no one pays taxes when most people do. In this case, presenting the descriptive norm of how frequent tax payments indeed are can help improve individual behavior.
Prescriptive norms	They refer to what society approves or disapproves of (i.e., what is considered to be right or wrong), regardless of how individuals actually behave. These norms are useful for reaffirming or encouraging positive individual behaviors while discouraging negative ones.
Reciprocity	It is a social norm that involves exchanges between individuals, in which one person responds to the action of another with a similar one. It most commonly refers to exchanges involving positive actions, like returning a favor with another similar favor. However, it can also refer to negative actions, such as punishing others in response to being the recipient of a harmful action.
Heuristics (rules of thumb)	They are recommendations that can help make decisions easier by offering mental shortcuts, or rules of thumb, to help guide the decision-making process.

Source: Martínez Villarreal et al. (2020)

Table 15 Mapping nudges to biases

Nudges for types of barriers (Examples)	**Increase salience of information**	**Framing**	**Commitment contracts**	**Social incentives and peer effects**	**Lottery and financial incentives**	**Self-monitoring, feedback, reminders**	**Channel factors**
Bounded rationality							
Optimism and overconfidence (type I) *I am aware of the issue, but I am confident this does not or will not happen to me*	■						
Optimism and overconfidence (type II) *If I make a commitment, I am sure I can achieve the goal I have a good chance of winning a lottery*			■		■		
Limited attention, vividness, availability heuristic, and priming *I base my choice on the information that is readily available in the moment*	■	■				■	■
Commitment consistency principle *I do not like to break a previous commitment*			■				
Loss and regret aversion *I really do not like to give up something that is mine (or that I may win) or that I am earning*					■		

(continued)

Table 15 (continued)

Status quo bias *I tend to choose the default option*		■					
Anchoring *I choose the normal option (middle of range)*		■					
Affect heuristic *I am likely to react to information if it is presented in an emotional context*	■						
Bounded willpower							
Time inconsistent preferences, hyperbolic discounting, present bias *I am aware of the problem and I want to do something about it, but I will do it tomorrow*		■	■		■		
Ego depletion and decision fatigue *I do not realize that I am not doing anything about the problem (I forget/my willpower has too many conflicting tasks to perform)*		■	■	■		■	■
Bounded selfishness							
Social norms I am fine because I am like others around me				■			

Source: Pinto et al (2014)

Source: Pinto et al. (2014)

References

Altinbasak-Farina, I., & Burnaz, S. (Eds.). (2019). *Ethics, social responsibility and sustainability in marketing*. Springer. https://doi.org/10.1007/978-981-13-7924-6

Alves, H., & Galan-Ladero, M. M. (2019). Theoretical background: Introduction to social marketing. In M. M. Galan-Ladero & H. Alves (Eds.), *Case studies on social marketing. A global perspective*. Springer.

Alves, H., Galan-Ladero, M. M., & Rivera, R. G. (2021). Theoretical background: Social marketing and quality of life. In M. M. Galan-Ladero & R. Rivera (Eds.), *Applied social marketing and quality of life. Applying quality of life research series*. Springer.

Alves, H., & Vazquez, J. L. (Eds.). (2013). *Best practices in marketing and their impact on quality of life*. Springer. https://www.springer.com/gp/book/9789400758773

Acquier, A., Valiorgue, B., & Daudigeos, T. (2017). Sharing the shared value: A transaction cost perspective on strategic CSR policies in global value chains. *Journal of Business Ethics, 144*, 139–152.

Basil, D. Z., Diaz-Meneses, G., & Basil, M. D. (2019). *Social marketing in action. Cases from around the world*. Springer.

Belz, F. M., Peattie, K., & Galí, J. M. (2013). *Marketing de Sostenibilidad*. Profit Editorial.

Berle, A. A. (1932). For whom corporate managers are trustees: A note. *Harvard Law Review, 45*(8), 1365–1372.

Bowen, H. R. (1953). *Social responsibilities of the Businessman*. Harper & Row.

Campbell, J. L. (2007). Why would corporations behave in socially responsible ways? An institutional theory of corporate social responsibility. *Academy of Management Review, 32*(3), 946–967.

Carroll, A. B. (1979). A three-dimensional conceptual model of corporate performance. *Academy of Management Review, 4*(4), 497–505.

Carroll, A. B. (1991). The pyramid of corporate social responsibility: Toward the moral management of organizational stakeholders. *Business Horizons, 34*(4), 39–48.

Coase, R. H. (1937). The nature of the firm. *Economica*, New Series, *4*(16), 386–405. https://doi.org/10.1111/j.1468-0335.1937.tb00002.x

Coase, R. H. (1992). The institutional structure of production. *The American Economic Review, 82*(4), 1–16. https://doi.org/10.1515/jeeh-1991-0401

Crowther, D., & Seifi, S. (Eds.). (2021). *The Palgrave handbook of corporate social responsibility*. Palgrave Macmillan.

Dahlsrud, A. (2008). How corporate social responsibility is defined: An analysis of 37 definitions. *Corporate Social Responsibility and Environmental Management, 15*(1), 1–13.

Daub, C. H., & Ergenzinger, R. (2005). Enabling sustainable management through a new multidisciplinary concept of customer satisfaction. *European Journal of Marketing, 39*(9/10), 998–1012.

Davis, K. (1960). Can business afford to ignore corporate social responsibilities? *California Management Review, 2*, 70–76.

Dodd, E. M., Jr. (1931). For whom are corporate managers trustees. *Harvard Law Review, 45*, 1145.

Donovan, R., & Henley, N. (2010). *Principles and practice of social marketing: An international perspective*. Cambridge University Press.

Dueñas-Ocampo, S., Perdomo-Ortiz, J., & Villa-Castaño, L. E. (2014). El concepto de consumo socialmente responsable y su medición. Una revisión de la literatura. *Estudios Gerenciales: Journal of Management and Economics for Iberoamérica, 30*(132), 287–300.

Elkington, J. (1997). *Cannibals with forks: Triple bottom line of 21st century business*. Capstone Publishing Limited.

European Commission. (2011). A renewed EU strategy 2011–14 for corporate social responsibility. Communication from the Commission to the European Parliament, the Council, the European Economic and Social Committee and the Committee of the Regions. /* COM/2011/0681 final */ Document 52011DC0681

Evans, J. S. B. (2008). Dual-processing accounts of reasoning, judgment, and social cognition. *Annual Review of Psychology, 59*, 255–278.
Frederick, W. C. (2008). *Corporate social responsibility: Deep roots, flourishing growth, promising future*. Oxford University Press.
Freeman, R. E. (1984). *Strategic management: A stakeholder approach*. Pitman.
Freeman, R. E., Harrison, J. S., Wicks, A. C., Parmar, B. L., & De Colle, S. (2010). *Stakeholder theory. The state of the art*. Cambridge University Press.
Frederick, W. C. (1960). The growing concern over business responsibility. *California Management Review, 2*(4), 54–61.
French, J., & Gordon, R. (2020). *Strategic social marketing: For behaviour and social change* (2nd ed.). SAGE Publications, Inc.
Friedman, M. (1962). *Capitalism and freedom*. University of Chicago Press.
Friedman, M. (1970). *The social responsibility of business is to increase its profits* (pp. 32–33). New York Times Magazine.
Galan-Ladero, M. M., & Alves, H. (2019). *Case studies on social marketing. A global perspective*. Springer.
Galan-Ladero, M. M., & Alves, H. (2023). *Social Marketing and Sustainable Development Goals (SDGs). Case Studies for a Global Perspective*. Springer Business Cases. Springer.
Galan-Ladero, M. M., Galera-Casquet, C., & Alves, H. M. (Eds.). (2021). *Cause-related marketing. Classroom companion: Business*. Springer. https://doi.org/10.1007/978-3-030-65455-9
Garriga, E., & Melé, D. (2004). Corporate social responsibility theories: Mapping the territory. *Journal of Business Ethics, 53*(1), 51–71.
Gautier, A., & Pache, A. C. (2015). Research on corporate philanthropy: A review and assessment. *Journal of Business Ethics, 126*, 343–369.
Haski-Leventhal, D., Kach, A., & Pournader, M. (2019). Employee need satisfaction and positive workplace outcomes: The role of corporate volunteering. *Nonprofit and Voluntary Sector Quarterly, 48*(3), 593–615.
Johnson, H. L. (1971). *Business in contemporary society: Framework and issues*. Wadsworth.
Jones, T. M. (1980). Corporate social responsibility revisited, redefined. *California Management Review, 22*(3), 59–67.
Kahneman, D. (2003). Maps of bounded rationality: Psychology for behavioral economics. *American Economic Review, 93*(5), 1449–1475.
Kakabadse, N. K., Rozuel, C., & Lee-Davies, L. (2005). Corporate social responsibility and stakeholder approach: A conceptual review. *International Journal of Business Governance and Ethics, 1*(4), 277–302.
Kao, T. Y., Chen, J. C., Wu, J. T. B., & Yang, M. H. (2016). Poverty reduction through empowerment for sustainable development: A proactive strategy of corporate social responsibility. *Corporate Social Responsibility and Environmental Management, 23*(3), 140–149.
Kotler, P., Hessekiel, D., & Lee, N. R. (2012). *Good works: Marketing and corporate initiatives that build a better world and the bottom line*. Wiley.
Kotler, P., Kartajaya, H., & Setiawan, I. (2011). *Marketing 3.0*. LID Editorial.
Kotler, P., & Lee, N. (2005). *Corporate social responsibility*. Wiley.
Kotler, P., & Lee, N. (2008). *Social marketing influencing behaviors for good* (3rd ed.). SAGE Publications.
Kotler, P., & Zaltman, G. (1971). Social marketing: An approach to planned social change. *Journal of Marketing, 35*(July), 3–12.
Llopis Goig, R. (2009). Consumo responsable y globalización reflexiva: Un estudio referido al comercio justo en España. *Revista Española del Tercer Sector, 11*, 145–165.
Martínez Villarreal, D., Rojas Méndez, A. M. & Scartascini, C. (2020). *Behavioral economics can help fight coronavirus*. Inter-American Development Bank. https://doi.org/10.18235/0002315.
McWilliams, A., & Siegel, D. (2001). Corporate social responsibility: A theory of the firm perspective. *Academy of Management Review, 26*(1), 117–127.

Morrós-Rivera, J., & Vidal-Martínez, I. (2005). *La Responsabilidad Social Corporativa*. Editorial Fundación Confemetal.
Nave, A., & Ferreira, J. (2019). Corporate social responsibility strategies: Past research and future challenges. *Corporate Social Responsibility and Environmental Management, 26*(4), 885–901. https://doi.org/10.1002/csr.1729
Ogaki, M., & Tanaka, S. C. (2017). Behavioral economics, toward a new economics by integration with traditional economics. *Springer Texts in Business and Economics*. https://doi.org/10.1007/978-981-10-6439-5_1
Peattie, K. (1992). *Green marketing*. Pitman Publishing.
Peattie, K. (1995). *Environmental marketing management: Meeting the green challenge*. Pitman Publishing.
Penelas-Leguía, A., Galera-Casquet, C., Galan-Ladero, M., & Valero-Amaro, V. (2012). *Marketing Solidario. El Marketing en las Organizaciones no Lucrativas*. Pirámide.
Pinto, D. M., Ibarrarán, P., & Stampini, M. (2014). *Applying behavioral tools to the design of health projects*. Policy Brief No. IDB-PB-228, July, Inter-American Development Bank. https://publications.iadb.org/en/applying-behavioral-tools-design-health-projects
Porter, M. E., & Kramer, M. R. (2006). The link between competitive advantage and corporate social responsibility. *Harvard Business Review, 84*(12), 78–92.
Samson, A. (Ed.). (2014). The Behavioral Economics Guide 2014 (with a foreword by George Loewenstein and Rory Sutherland) (1st ed.). Retrieved from http://www.behavioraleconomics.com
Santesmases, M. (1999). *Marketing. Conceptos y Estrategias*. 4th Edition. Piramide.
Santesmases, M. (2007). *Marketing. Conceptos y Estrategias*. 5th revised Edition. Piramide.
Ulusoy, E. (2016). Experiential responsible consumption. *Journal of Business Research, 69*(1), 284–297.
Thaler, R. H., & Sunstein, C. R. (2008). *Nudge: Improving decisions about health, wealth, and happiness*. Penguin.
The United Nations. (2015). Resolution adopted by the General Assembly on 25 September 2015 "Transforming Our World: the 2030 Agenda for Sustainable Development". Accessed October 11, 2022, from https://www.un.org/en/development/desa/population/migration/generalassembly/docs/globalcompact/A_RES_70_1_E.pdf
The United Nations. (2019). *The Lazy Person's guide to saving the world*. Accessed October 11, 2022, from https://www.un.org/sustainabledevelopment/takeaction/
The United Nations. (2022a). *Goal 12: Ensure sustainable consumption and production patterns*. Accessed October 11, 2022, from https://www.un.org/sustainabledevelopment/sustainable-consumption-production/
The United Nations. (2022b). *17 Goals to transform Our world*. Objetivos de Desarrollo Sostenible. Accessed October 11, 2022, from https://www.un.org/sustainabledevelopment/
The United Nations Environment Programme, One Earth. (2018). *Sustainable lifestyles: Options and opportunities*. UN Environment.
Vázquez, J. L., & Lanero, A. (2021). Consumer transition to a green economy: The role of third-party certified eco-labels. *SHS Web of Conferences, 120*, 02002.
Vredenburg, J., Kapitan, S., Spry, A., & Kemper, J. A. (2020). Brands taking a stand: Authentic brand activism or woke washing? *Journal of Public Policy & Marketing, 39*(4), 444–460.
Warde, P. (2018). *The invention of sustainability: Nature and Destiny, c. 1500–1870*. Cambridge University Press.
Warde, P., Robin, L., & Sörlin, S. (2018). *The environment: A history of the idea*. Johns Hopkins Press.

Wood, D. J. (1991). Corporate social performance revisited. *Academy of Management Review, 16*(4), 691–718.
World Bank. (2015). *World development report 2015: Mind, society, and behavior*. World Bank. https://doi.org/10.1596/978-1-4648-0344-4
World Commission on Environment and Development - WCED. (1987). *Our common future (the Brundtland report)*. Oxford University Press.
Zucker, L. G. (1987). Institutional theories of organization. *Annual Review of Sociology, 13*, 443–464. https://doi.org/10.1146/annurev.so.13.080187.002303

Part I

Sustainable Tourism and Fashion

Corporate Social Marketing, Drunken Tourism and Sustainability: The Case of Meliá Hotels International in Magaluf, Spain

Marco A. Cruz-Morato, Carla Rodriguez-Sanchez, and Carolina Sanchez

Learning Objectives

General objective: to learn about the role of corporate social marketing in changing tourists' behaviour for destination sustainability

Specific objective 1: to learn how corporate social marketing can help build brand equity

Specific objective 2: to learn how social marketing communication campaigns should be designed on social media

1 Introduction

One of the biggest problems with tourism today is the rapid rise in the number of visitors, which has huge negative impacts on the environment. According to the UNWTO (United Nations World Tourism Organization) (2019), by 2030, the number of international tourist arrivals worldwide will grow by an average of 3.3% per year. This growth equates to about 43 million people per year [1] compared to an average increase of 28 million people per year in the period 1995–2010. This massive arrival of tourists can be an opportunity. However, it can also be a threat if

[1] See Frey and Briviba (2021) and Hall (2014).

M. A. Cruz-Morato
Faculty of Tourism, University of Malaga, Malaga, Spain
e-mail: mcruz@uma.es

C. Rodriguez-Sanchez (✉)
Faculty of Economic and Business Sciences, University of Alicante, Alicante, Spain
e-mail: carla.rodriguez@ua.es

C. Sanchez
Faculty of Economic Sciences, National University of Cordoba, Córdoba, Argentina
e-mail: csanchez@unc.edu.ar

E. C. Bianchi et al. (eds.), *Responsible Consumption and Sustainability*, Springer Business Cases, https://doi.org/10.1007/978-3-031-30742-3_2

its potential negative impacts are not controlled. Hence, sustainable tourism must be encouraged. Social marketing can be a powerful tool to achieve this aim because social marketing strategies encourage the responsible behaviour of visitors (Torres-Moraga et al., 2021), creating a better balance between the interests of tourists, residents and local businesses (Hall, 2014).

Corporate social marketing strategies and actions have been adopted by some hotels to encourage the sustainable behaviour of their customers.[2] A prime example is *Meliá Hotels International* in the tourist town of Magaluf in Spain. This hotel group modernised its hotels to offer premium facilities and services. It thus distanced itself from the offer of low-cost tourism. At the same time, it developed corporate social marketing strategies to attract a more responsible clientele and convince the less responsible tourist segment that it was better to enjoy more peaceful and relaxing holidays.

The aim of this *case study* is to describe the corporate social marketing actions carried out by *Meliá Hotels International* in the tourist town of Magaluf in Spain. These actions were supported by complementary initiatives by local public authorities aimed at changing patterns of behaviour and promoting sustainable tourism. The case provides impact analysis of social marketing campaigns, focusing on online traffic. Content analysis is also provided.

2 Development of the Case

2.1 Context

2.1.1 Magaluf and Drunken Tourism

Magaluf is a coastal tourist town on the island of Mallorca (Balearic Islands, Spain). Its population is around 5,000 inhabitants (Ibestat, 2021). From the 1960s onwards, the tourist boom has made Magaluf one of the main (mass) tourist destinations on the island. Most tourists are young people from the United Kingdom looking for a budget destination with beaches, parties and cheap alcohol. Over the years, the town has gained international fame as a low-cost party tourist destination.

The image problems date back to the year 2000, when sex scandals linked to alcohol, drugs and debauchery among young tourists came to light. To make the situation worse, *balconing*, an activity that consists of jumping into the hotel pool from the balcony of hotel rooms located on the upper floors of buildings, began to claim the lives of many young people. These scandals have stained the international image of the destination of Magaluf. They have also caused great discomfort and insecurity among its inhabitants. This situation has attracted complaints from the family tourism segment. The reasons for Magaluf's situation can be summarised as follows (Prieto, 2020):

[2] See, for example, González-Rodríguez et al. (2019).

1. The behaviour of tour operators, who seek to sell as many holiday packages as possible at the lowest possible price, has meant that the number of tourists has been prioritised over quality.
2. The hyper-specialisation of the area has been the low-cost British market, which has led to a "Britanisation" of Magaluf.
3. The massive dissemination of news and reports through the media (written press, television and social media) has encouraged sensationalism and exaggerated reports to gain an audience.

2.1.2 Meliá Hotels International

One of the leaders of corporate social marketing actions in Magaluf has been *Meliá Hotels International*. This hotel group was founded in 1956 in Palma de Mallorca (Spain). It has 11 tourist accommodations in this area. Some of the most prestigious are *Meliá South Beach*, *Meliá Calviá Beach*, *Innside Cala Blanca*, *Sol Wave House*, *Sol Barbados* and *Sol Guadalupe*.

In 2011, *Meliá Hotels International* started the "Sol Calvià Resort" project. Supported by the local authorities, they decided to invest heavily in the modernisation of the tourist destination of Magaluf. Following the success stories of Miami Beach and Barcelona, their idea was to achieve a total overhaul in the Magaluf area. The project involved the remodelling of six of their hotels in addition to urban planning modifications such as the creation of a promenade, the widening of pavements, the modernisation of shopping areas and the construction of parking lots with new entertainment and food offerings. The main aim was to modernise Magaluf by providing a top-quality offering.

In 2015, "The New Magaluf" was presented to the world at the London Tourism Fair (most of the intermediaries were British). *Meliá Hotels International*, together with nine other private companies in the Mallorca tourism sector, representatives from the Balearic Government and the Calvià City Council, announced their goals for the regeneration of the destination of Magaluf. Both the City Council of Calvià and the Balearic Government showed commitment and interest in changing the tourism model to seek a more sustainable form of tourism that would benefit the town. This new form of tourism was high-quality tourism that focused on tourism excellence as opposed to mass or low-cost tourism. It meant changing the tourism model to benefit the destination by making it more sustainable. With this project, they facilitated and streamlined the procedures of companies committed to change.

Following the example of *Meliá Hotels International*, several hotels and companies joined the Magaluf regeneration project. The hotel offering was increased, and the level of quality was raised through renovations and the expansion of services. In August 2018, 60% of the hotel offering of Magaluf was four or five stars. *Meliá Hotels International* repositioned its hotels and changed its customer segment, starting in 2013. In 2018, 70% of its customers were over 30 years old, with 42.7% staying at the hotels as a couple and 30.1% as a family. The total investment by *Meliá Hotels International* in this Magaluf regeneration project was around 250 million euros (Ruiz Collado, 2018). However, they still needed to explore their communication to attract more respectful guests and directly or indirectly influence the behaviour of visitors who continued to behave uncivilly.

Since 2019, with the relevant changes complete, *Meliá Hotels International* has run the *#TheNewMagaluf* campaign. The thinking behind this campaign is, what use is an investment of 250 million euros if the viral news about the destination of Magaluf continues to be about drunken tourism and balconing? Meliá created this campaign and promoted it on the company's official social media accounts through LinkedIn, Facebook, Instagram and Twitter. The campaign was accompanied by a video.[3] The target audience consisted of both customers (intermediaries and end customers, mainly British) and the residents of the area.

The public authorities, led by the City Council of Calvià, also supported this communication strategy. They developed social marketing campaigns, mainly aimed at visitors, that focused on the health and safety problems caused by inappropriate behaviour versus the advantages of responsible leisure. They sought not only to attract a more responsible and sustainability-oriented clientele but also to convince the less responsible segment that it was better to enjoy more relaxing holidays, avoiding excessive alcohol consumption and uncivil behaviour. This approach is along the same lines as the *#TheNewMagaluf* campaign. All these communication campaigns by both private and public actors were carried out through traditional and social media.[4]

2.2 Analysis and Results

Meliá Hotels International hired a digital marketing agency to analyse the results of the online communication actions carried out in the *#TheNewMagaluf* campaign. Content analysis of these social marketing communication actions was performed, focusing on the following aspects: message, tone, purpose, graphic design, means of dissemination and target audience.[5] Analysis of their impact[6] was also conducted, focusing on online traffic in the form of YouTube views and impact on social media. Where possible, the analysis examined the impact using *Hashtagify*. The analysis studied variables such as popularity, total publications, recent interactions, related hashtags and others.

2.2.1 Content Analysis

Regarding the content analysis of the *#TheNewMagaluf* campaign, the spot has a very professional tone. It is also informative and positive. It uses background

[3]The video can be viewed at https://www.ultimahora.es/noticias/part-forana/2019/07/30/1097599/melia-lanza-campana-para-dar-conocer-the-new-magaluf.html.

[4]Traditional media are not the target of this case. This case study focuses on online corporate social marketing campaigns.

[5]Following Cruz-Morato and Dueñas-Zambrana (2021).

[6]See, for example, Hays et al. (2013).

chill-out music and an elegant graphic design focused on relaxing activities, nature, responsible fun and service quality.

Regarding the message and graphic design, some scenes at the beginning were recorded with a drone. These scenes focus on the natural surroundings, beaches, swimming pools and hotel building. The Meliá brand can be clearly seen under the caption, "new experiences in hotels", reinforcing the idea that the tourist offering has changed. It is a new concept, contrary to "what you know about Magaluf".

Next, the video shows scenes of different hotel services such as the restaurant, the pool (with the image of an open book on a table to represent the peace and relaxation of the New Magaluf), a bartender preparing a cocktail and scenes of cooking and food. It then shows a *DJ* and a group of people aged 25–30 years dancing to the caption "cosmopolitan atmosphere". The new positioning cannot completely reject all forms of nightlife; responsible party tourism is also encouraged.

The sea and sunset are then shown. The images sell the "Mediterranean lifestyle", with some people doing yoga, returning to the new *zen* and sustainable positioning. The video also shows the hotel and its swimming pools, the "cosy beach coves" with crystal clear waters to reinforce the value of nature and children running along the shore to the caption "entertainment for all". This image provides new evidence of the interest in attracting families to this new sustainable concept of Magaluf. Next, the spot shows some images of young people *surfing* to the caption "energy and social vibes" and again some children having fun at a theme park. Shopping scenes are also included.

Finally, the video shows young people aged about 25–30 years toasting with shots and having fun dancing at a concert. Although this image again depicts a party, the details are different from the idea of the massive uncontrolled parties associated with drunken tourism. It reflects a more sophisticated, exclusive and responsible form of leisure. At this point, the caption "*Meliá Hotels International* presents *#TheNewMagaluf*" appears. The beaches and crystal clear water are once again shown at the end of the spot, and the caption reads "have you heard?" with the name of the company.

As explained earlier, the aim of this campaign was to convince responsible tourists to visit Magaluf and inform regular tourists to promote the new Magaluf lifestyle and persuade them not to behave irresponsibly. Even if they are not represented by the new positioning or cannot afford the new high-quality services, the campaign would achieve the social goal of greater sustainability. Magaluf is presented as the best destination in the Mediterranean to combine peace, nature and sustainable tourism. It is depicted as a place where young people and families can have fun while enjoying responsible leisure activities in a sophisticated, exclusive, high-quality context.

This campaign was disseminated via the company's social media accounts on *LinkedIn*, *Facebook*, *Instagram* and *Twitter*. In addition to the main spot with a duration of two minutes (2'11"), a shorter version (31") was also produced. On the *Meliá Hotels International* website, there was also a section on *#TheNewMagaluf* with the following description: "Peace next to the Mediterranean, sophistication and

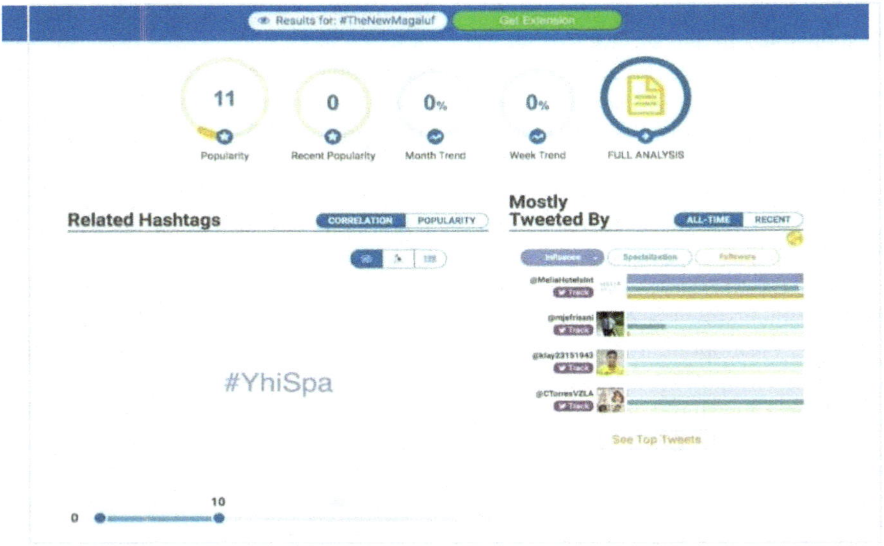

Fig. 1 Hashtag *#TheNewMagaluf* results on *Twitter*.[7] Source: Authors using data from Hashtagify (2022)

the best family entertainment. This is what happens in Magaluf, and you would never have imagined it...".

2.2.2 Analysis of the Impact of the Campaign

The impact of the campaign was also analysed using the hashtag *#TheNewMagaluf* on the most popular social media. Compared to the usual traffic figures for company (and sector) campaigns, the impact was small in terms of interactions and reach.

- On YouTube, the video received 4215 views.[7]
- On LinkedIn and Facebook, the campaign was shared mainly by Meliá hotel employees.

The impact on Twitter (Fig. 1) and Instagram (Figs. 2 and 3) was measured in *Hashtagify*.

On *Twitter* and *Instagram*, the impact was also quite low, with very few views and little engagement. However, on *Instagram*, the number of total posts and recent interactions (calculated based on the number of likes and comments on the last 15 posts using the hashtag) increased slightly between June and November 2021 and August 2022. It should be noted that with these hashtags, other completely separate companies from *Meliá Hotels International* also participated, such as the *Hotel*

[7] As of August 8, 2022.

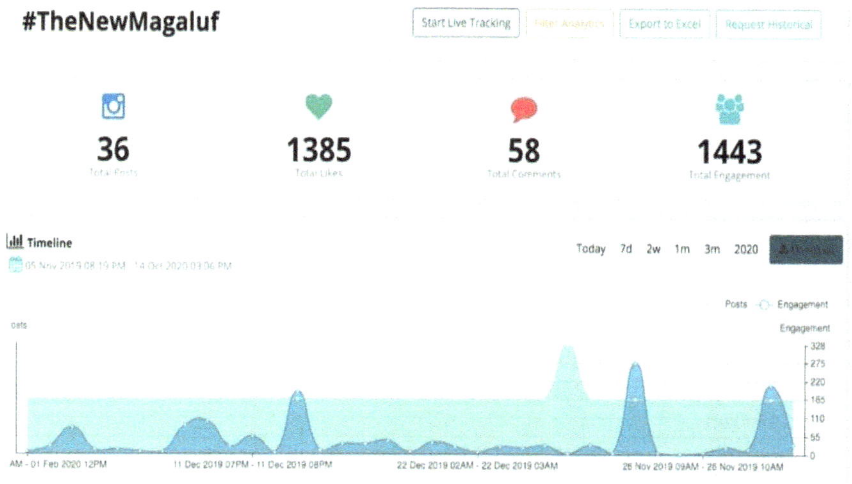

Fig. 2 Hashtag *#TheNewMagaluf* results on Instagram.[7] Source: Authors using data from Hashtagify (2022)

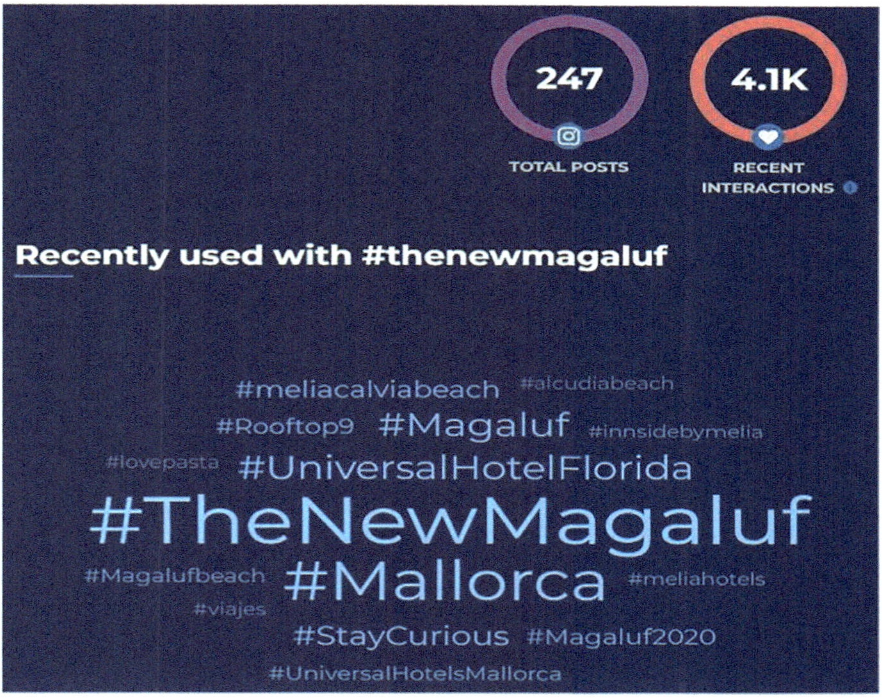

Fig. 3 Hashtag *#TheNewMagaluf* results on *Instagram.*[7] Source: Authors using data from Hashtagify (2022)

Florida, the *Folies Beach Club* restaurant and the *UrbanCleaner* hotel disinfection and cleaning company.

3 Conclusions

The corporate social marketing actions promoted by *Meliá Hotels International*, which were later supported by other private and public actors, involved the application of marketing tools such as digital communication to encourage behavioural changes for the benefit of companies and local residents. Strategies of target market segmentation and destination positioning were used. These actions led to changes in the offering of tourist services (product), price and communication actions (with a limited impact). In summary, the strategy had a positive impact in terms of the tourism sustainability of Magaluf, although the communication campaigns had a limited impact.

As an area of improvement, these campaigns should be designed with more persuasive elements that have been tested in the academic literature on the sustainable behaviour of tourists. For example, they could focus on personal benefits and social norms (Hardeman et al., 2017). In addition, these communication campaigns should also follow the main principles of communication in social marketing to achieve sustainable behavioural changes. For example, campaign effectiveness is greater when the message is consistent with all advocacy efforts and when it is in harmony with people's context.[8]

Finally, if the campaign is going to be published on social media, those responsible for the campaign must consider how these social media should be used to change people's behaviour more efficiently and effectively. Accordingly, the messages must be designed to attract the clientele. To do so, the focus must be not only on designing attractive messages by offering information but also on achieving interaction with the audience and developing bilateral communication.[9]

The results of the social marketing actions analysed in this *case study* can offer a starting point to leave behind the image of Magaluf as a destination for drunken and low-cost tourism. There is still the challenge of continuing to reposition the destination in the post-pandemic age. Finally, this case study highlights the importance of public–private partnerships in promoting behavioural change.

[8] See, for example, Pearson and Perera (2018).

[9] See, for example, Shawky et al. (2019).

4 Discussion Questions

Question 1. This case study shows that conducting corporate social marketing actions is crucial to improve the brand equity of a tourist destination (in this case, Magaluf). How do you think corporate social marketing activities can help build the brand equity of a tourist destination?

Question 2. This case study shows that the *#TheNewMagaluf* communication campaign on social media had little impact among the public. How should this type of campaign be designed to be more effective?

Question 3. Considering the results of the campaign presented in this case study and the effects of the COVID-19 pandemic on the tourism sector, design a new *#TheNewMagaluf* communication campaign, assuming that you are the marketing consultant for the Meliá group. Use the Social Media for Behaviour Change (SM4BC) toolkit,[10] explained in the previous point, as a starting point to consider the two main recipients of the campaign: (1) tourists with a family profile who seek rest and relaxation and (2) young tourists who visit the destination in search of fun.

Appendix: Teaching Notes for the Case

Synopsis

This case study uses a corporate social marketing perspective to present an example of tourism sustainability promotion. It focuses on the case of Magaluf (Spain), a tourist destination known for its history of drunken tourism and balconing. This case study examines the actions of *Meliá Hotels International*, in particular the *#TheNewMagaluf* campaign. The campaign content was analysed. The analysis focused on the message, tone, purpose, graphic design, media and target audience. The impact of the online social marketing campaigns (mainly in terms of online traffic through views on YouTube and social media impact) was also analysed using *Hashtagify*. These social marketing actions, among others, have enhanced tourism sustainability in Magaluf and have created benefits for both companies and residents. They provide an opportunity to reposition the image of Magaluf.

Potential Audience and Instructor's Material

The case study is designed to be used in marketing management both in courses at the MBA and Executive Programme level and in advanced degree level courses. The case study exemplifies corporate social marketing campaigns focused on the

[10] Social Media for Behaviour Change (SM4BC) toolkit: https://preparecenter.org/site/sm4bc-toolkit/

promotion of sustainable tourism that use traditional marketing resources such as positioning strategy, branding and marketing communication.

The case study highlights the goal of a social marketing campaign, namely, to create benefits for the community as a whole. Specifically, business and government formed an alliance so that Magaluf could reposition itself as a peaceful, high-quality tourist destination for families, distancing itself from its former image as a destination of drunkenness and disorder. The construction of the new brand of New Magaluf will benefit the community of Magaluf while promoting local tourism growth. The community of Magaluf can coexist in harmony with tourists and local citizens, while local tourism can grow, thanks to the arrival of new visitors with a different socioeconomic and behavioural profile.

Learning Objectives

General objective: to learn about the role of corporate social marketing in changing tourists' behaviour for destination sustainability

Specific objective 1: to learn how corporate social marketing can help build brand equity

Specific objective 2: to learn how social marketing communication campaigns should be designed on social media

Time Frame for Class Discussion

This case can be taught in one session of 60–100 minutes. The duration will depend on the number of students participating in the discussion activities and the preparatory work by students before the class.

Suggested Discussion Questions

Question 1. This case study shows that conducting corporate social marketing actions is crucial to improve the brand equity of a tourist destination (in this case, Magaluf). How do you think corporate social marketing activities can help build the brand equity of a tourist destination?

Question 2. This case study shows that the *#TheNewMagaluf* communication campaign on social media had little impact among the public. How should this type of campaign be designed to be more effective?

Question 3. Considering the results of the campaign presented in this case study and the effects of the COVID-19 pandemic on the tourism sector, design a new *#TheNewMagaluf* communication campaign, assuming that you are the marketing consultant for the Meliá group. Use the Social Media for Behaviour Change

(SM4BC) toolkit,[11] explained in the previous point, as a starting point to consider the two main recipients of the campaign: (1) tourists with a family profile who seek rest and relaxation and (2) young tourists who visit the destination in search of fun.

Analysis

Answer to Question 1. *This case study shows that conducting corporate social marketing actions is crucial to improve the brand equity of a tourist destination (in this case, Magaluf). How do you think corporate social marketing activities can help build the brand equity of a tourist destination?*

Numerous articles explain the role of corporate social marketing in creating brand equity. One of the most important and well-cited articles is by Hoeffler and Keller (2002). They highlight the following six means through which corporate social marketing programmes can help build brand equity:

1. Build brand awareness.
2. Improve brand image.
3. Establish brand credibility.
4. Evoke brand feelings.
5. Create a sense of brand community.
6. Gain commitment to the brand.

Considering this theoretical framework, to build the brand equity of Magaluf, it is crucial to generate brand awareness. This aim involves getting tourists and tour operators to recognise the Magaluf brand. They must associate it with its new attributes of being the best destination in the Mediterranean, combining peace and nature, where young people and families can have fun and enjoy responsible leisure activities in a sophisticated, exclusive and high-quality context. They must also define a positioning strategy based on these attributes.

The suggested actions are as follows:

- Advertising campaigns should be carried out through different offline and online media, where the attributes of the destination are presented. Doing so will help the audience associate the name of the destination with images, products and services related to its new attributes. The campaigns must be segmented to target different target audiences (e.g. young segments and the family segment), with a focus on the attributes that are most valued by each group of tourists. Then, the means of communication can be chosen. Likewise, messages that appeal to the emotions

[11] Social Media for Behaviour Change (SM4BC) toolkit: https://preparecenter.org/site/sm4bc-toolkit/

and feelings that the destination awakens should be used, especially to elicit feelings of rest, holidays and enjoyment of natural and relaxing environments.

- In the communication and dissemination actions, the image of the brand should be promoted through famous personalities who convey the desired brand image, associating the chosen values, beliefs and lifestyles with the tourist destination. There must be a strong fit between the image of the destination and the selected public personality.
- Tour operators and local government agencies associated with tourism must participate in fairs, conferences and events that share and promote the new image of Magaluf. They must present to the public their renewed offer of tourist services, highlighting the premium services that are now provided.
- Tour operators and local government agencies associated with tourism should sponsor events where responsible and sustainable tourism is promoted, as well as other events (e.g. sports, social and cultural) with values linked to the destination of Magaluf.
- Another proposal is to develop activities, events and actions that respond to the Sustainable Development Goals related to tourism: Goal 8 (Decent Work and Economic Growth), Goal 12 (Responsible Consumption and Production) and Goal 14 (Life Below Water). Mail and communication actions should raise the visibility of Magaluf's commitment to the SDGs.
- Generating brand credibility is essential for Magaluf's new image to be reliable. Messages must demonstrate Magaluf's expertise in providing high-quality tourism services and products. Thus, tourism agencies and operators, as well as government agencies related to tourism, should obtain ISO standards for tourism services and exhibit the corresponding certifications. Examples include ISO 21902, ISO 18513 and ISO 18065.
- Social marketing campaigns should be implemented to avoid excessive alcohol consumption. Such campaigns include de-incentivising the excessive consumption of alcoholic beverages. Appeals to social norms, self-respect and self-care can persuade people to avoid excessive consumption. Likewise, events, activities and actions that promote healthy fun and rest can be included in the campaigns.
- A sense of community with the Magaluf brand can be created. Digital communication actions can help promote a sense of community by transmitting the values of the tourist destination and associating it with a cause or theme such as care for the environment, health and well-being, healthy fun or others. To define the cause, research should be carried out to identify the profile of tourists who visit the destination to understand the topics and values with which they identify.

Answer to Question 2. *This case study shows that the #TheNewMagaluf communication campaign on social media had little impact among the public. How should this type of campaign be designed to be more effective?*

The recently published Social Media for Behaviour Change (SM4BC) toolkit is designed for use by anyone who wants to use social media to achieve positive behavioural change. The kit contains nine steps to design a campaign:[12]

1. *Diagnosis:* Know the internal capacity of the organisation to design and implement a campaign. This step involves describing the organisational mission, understanding the current use of social media, analysing whether social media fit with the organisational communication strategy, learning how communication with different stakeholders are coordinated and analysing whether the current approach to social media is effective.
2. *Set goals and objectives*: Define what you want to achieve. It is important to identify the problem, the factors that contribute to the problem and the factors that can be addressed through social media (because not all factors can be solved through the SM4BC process). The plan then goes on to define the communication goals for social media (i.e. how social media activities will contribute to behavioural change).
3. *Define the priority audience*: Decide who we need to reach. Identify what they know about the problem and what factors motivate them to change behaviours. Identify the barriers that inhibit such change. Identify who has the most influence over the priority audience (peer groups, influencers, opinion leaders, etc.). Understand the behaviour of the priority audience on social media (what they use, how, when, for what, etc.).
4. *Behaviour selection*: Define what we want the audience to do to address the problem. List all possible behaviours and then choose priority behaviours.
5. *Identify behavioural change mechanisms to help generate those changes*: Identify and eliminate barriers to behavioural change. Identify and promote the factors that motivate behavioural change. Identify the social factors that influence the adoption of the promoted behaviours.
6. *Define a messaging strategy*: Decide what should be communicated. This strategy includes communication, content, media, recall strategies, etc.
7. *Content production plan*. Decide how to create content. Define which social media will be used. Decide whether the publications will contain text, images, videos, graphics, etc. Plan the use of the contents.
8. *Community participation plan*: Define how to engage the community, how to interact with the audience on social media, how to track community engagement and how to tailor content for specific audiences.
9. *Evaluation of results*. Establish how we can know if we have been successful. Define metrics to measure the success of the campaign. Identify how to link social media activities to behavioural change. Design a research strategy to collect data to evaluate actions. Apply the results to improve and refine actions.

[12]Social Media for Behaviour Change (SM4CB) toolkit: https://preparecenter.org/site/sm4bc-toolkit/

Based on these steps, the actions by *Meliá Hotels International* in Magaluf for a new digital communication campaign for *#TheNewMagaluf* can be specified as follows:

1. *Diagnosis:* Because the initial campaign did not have a major impact, for the new campaign, you must first designate a person who is responsible for the campaign. You must then assemble a work team in charge of implementing the campaign. Perform an internal audit to obtain an internal marketing diagnosis.
2. *Set goals and objectives*: Define what you want to achieve. The problems addressed by the campaign are (1) how to re-position the image of Magaluf by presenting the new attributes and values of the destination and (2) how to change behaviours of excessive alcohol consumption, irresponsibility and balconing. The communication objectives for social media should have measurable, feasible and quantifiable objectives. The main ones should be number of organic visits to the page, number of followers, number of visits, bounce rate, volume of interactions and social traffic.
3. *Define the priority audience*: The priority audience consists of the target tourist segments of the destination of Magaluf. Identify and describe these segments. Previously, you will need to carry out research to answer the following questions: What do they know about the problem? What factors motivate them to change behaviours? What are the barriers that inhibit such change? Who has the most influence over the priority audience? How does the priority audience behave on social media: What do they use, how, when, for what, etc.?
4. *Behaviour selection*: Decide what we want the audience to do to address the problem. Based on the results from step 3, draw up a list of all possible behaviours, and then choose the priority behaviours.
5. *Identify behavioural change mechanisms to help generate those changes*: Identify and eliminate barriers to behavioural change. Identify and promote the factors that motivate behavioural change. Identify the social factors that influence the adoption of the promoted behaviours.
6. *Define a messaging strategy*: Communication, content, media, recall strategies, etc. are included. Include persuasive communication strategies that appeal to social norms, emotions, responsibility and values that can influence the behaviours of the audience.
7. *Content production plan.* Define which social media are going to be used (Facebook, Instagram, Twitter, TikTok, etc.) taking into account the target audience. Hire a digital marketing specialist to design the publications, which must contain text, images, videos, graphics, etc. Plan the content publication schedule.
8. *Community participation plan.* Design an action plan to get tourists in Magaluf to interact with the brand's activities on social media. The citizens of Magaluf could also be targeted. It is important to achieve two-way communication and to get users to feel heard and represented in the digital strategy.
9. *Evaluation of results.* Develop metrics to measure the success of the campaign in terms of the social media themselves (interactions, comments, shared messages,

etc.). To link social media activity with the behavioural change, the following should be considered: statistics from the local police (number of incidents, complaints, etc. related to excessive alcohol consumption or irresponsible behaviour by tourists), surveys to tourists and tour operators to gather data on the perception of safety, the behaviour of tourists in the destination, satisfaction with the destination, loyalty and recommendations. Apply the results to improve and refine activities.

Answer to Question 3. *Considering the results of the campaign presented in this case study and the effects of the COVID-19 pandemic on the tourism sector, design a new #TheNewMagaluf communication campaign, assuming that you are the marketing consultant for the Meliá group. Use the Social Media Toolkit for Behaviour Change (SM4BC) , explained in the previous point, as a starting point to consider the two main recipients of the campaign: (1) tourists with a family profile who seek rest and relaxation and (2) young tourists who visit the destination in search of fun.*

Students should perform desk research to learn about the current situation of the tourist destination in the post-pandemic era and the Meliá hotel chain. Make an overview of the situation.

Subsequently, design differentiated messages for each type of recipient. Emphasise the change of behaviour promoted among young tourists, discouraging excessive alcohol consumption and encouraging behaviour in keeping with the new profile of the destination based on peace, rest and fun without health risks.

References

Cruz-Morato, M. A., & Dueñas-Zambrana, C. (2021). Disability, labour inclusion and social Marketing in Spain: The ONCE foundation case. In M. M. Galan-Ladero & R. G. Rivera (Eds.), *Applied social marketing and quality of life*. Springer.

Frey, B. C., & Briviba, A. (2021). A policy proposal to deal with excessive cultural tourism. *European Planning Studies, 29*(4), 601–618.

González-Rodríguez, M. R., Martín-Samper, R. C., Köseoglu, M. A., & Okumus, F. (2019). Hotels' corporate social responsibility practices, organizational culture, firm reputation, and performance. *Journal of Sustainable Tourism, 27*(3), 398–419.

Hall, C. M. (2014). Framing behavioural approaches to understanding and governing sustainable tourism consumption: Beyond neoliberalism, "nudging" and "green growth"? *Journal of Sustainable Tourism, 21*(7), 1091–1109.

Hardeman, G., Font, X., & Nawijn, J. (2017). The power of persuasive communication to influence sustainable holiday choices: Appealing to self-benefits and norms. *Tourism Management, 59*, 484–493.

Hashtagify. (2022). *Search for hashtag popularity, trends and correlations (#TheNewMagaluf)*. Retrieved from: https://hashtagify.me/hashtag/thenewmagaluf. Accessed August 8, 2022.

Hays, S., Page, S. J., & Buhalis, D. (2013). Social media as a destination marketing tool: Its use by national tourism organisations. *Current Issues in Tourism, 16*(3), 211–239.

Hoeffler, S., & Keller, K. L. (2002). Building brand equity through corporate societal marketing. *Journal of Public Policy & Marketing, 21*(1), 78–89.

Ibestat. (2021). *Municipality in Cifras Calvià*. Retrieved from: https://ibestat.caib.es/ibestat/estadistiques/municipi-xifres/07011/Calvi%c3%a0. Accessed August 19, 2022.

Pearson, D., & Perera, A. (2018). Reducing food waste: A practitioner guide identifying requirements for an integrated social marketing communication campaign. *Social Marketing Quarterly, 24*(1), 45–57.

Prieto, C. (2020). Beyond the myth. Magaluf's rogue visionaries: you can travel to Spain with one euro. El Confidencial. Retrieved from: https://www.elconfidencial.com/cultura/2020-08- 09/magaluf-libro-turismo-historia_2708328/. Accessed August 19, 2022.

Ruiz Collado J. L. (2018). Meliá says it has managed to recover Magaluf with an innovative tourism project. Retrieved from: https://www.ultimahora.es/noticias/local/2018/07/06/1011597/melia-afirma-logrado-recuperate-magaluf-project-tourist-innovative.html. Accessed August 19, 2022.

Shawky, S., Kubacki, K., Dietrich, T., & Weaven, S. (2019). Using social media to create engagement: A social marketing review. *Journal of Social Marketing, 9*(2), 204–224.

Torres-Moraga, E., Rodriguez-Sanchez, C., & Sancho-Esper, F. (2021). Understanding tourist citizenship behavior at the destination level. *Journal of Hospitality and Tourism Management, 49*, 592–600.

World Tourism Organization (UNWTO). 2019). International Tourism Highlights, 2019 Edition. Retrieved from: https://www.unwto.org/publication/international-tourism-highlights-2019-edition. Accessed August 19, 2022.

Green Is in Fashion: Eco-Design and Circular Economy Strategies in the Fashion Industry

Antonio Chamorro-Mera and Rafael Robina-Ramírez

Learning Objectives

1. To reflect on the environmental implications involved in companies' buying, manufacturing, marketing and location decisions applied to the fashion and textile industry in this case
2. To encourage debate about the mass consumption society and marketing practices that encourage such, as well as the consumers' readiness to change their buying and consumer habits
3. To understand the circular economy concept by comparing it to the current linear economy
4. To understand the different ways a circular business model can be designed through eco-design and materials and waste management

1 Introduction

Open your wardrobe and take a look at the clothes you own. These clothes define your personality, your lifestyle and even your core values. But your clothes are also linked to your ecological footprint. The amount of clothing you own, the materials they are made from and where they were manufactured all have a greater environmental impact than we expect.

If we think about the whole value chain, that is, from growing or producing the raw materials (cotton and polyester, mainly) right through to their use and disposal as waste, the fashion sector has one of the most negative effects on the environment in terms of greenhouse gas (GHG) emissions, water and chemical use and wastewater production. This impact is exacerbated due to the mass consumption culture that

A. Chamorro-Mera (✉) · R. Robina-Ramírez
Universidad de Extremadura, Badajoz, Spain
e-mail: chamorro@unex.es; rrobina@unex.es

E. C. Bianchi et al. (eds.), *Responsible Consumption and Sustainability*, Springer Business Cases, https://doi.org/10.1007/978-3-031-30742-3_3

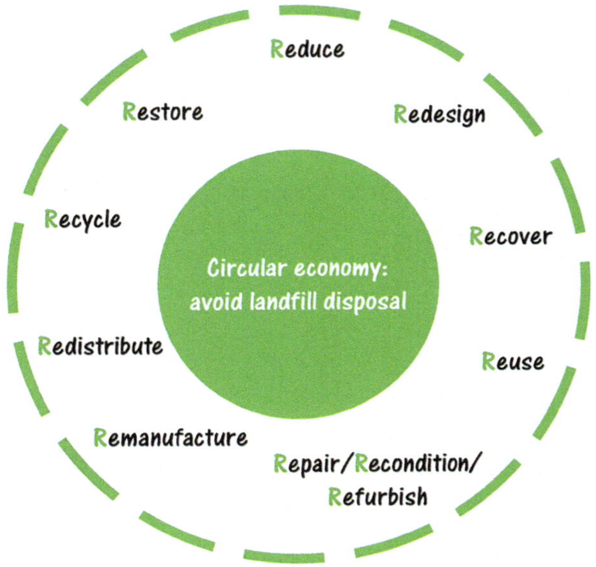

Fig. 1 The circular economy's multi-R approach. Source: own creation

dominates developed societies and has given rise to fast fashion: buying many clothes at low prices and discarding them after a short time.

The impact is of such magnitude that a change is urgently needed in the sector. It should move towards a circular economy model, understood as "a systems solution framework that tackles global challenges like climate change, biodiversity loss, waste, and pollution. It is based on three principles, driven by design: eliminate waste and pollution, circulate products and materials (at their highest value), and regenerate nature" (Ellen MacArthur Foundation, 2022). The circular economy reduces the use of natural resources and makes efficient use of them to extend products' life cycles. When a product reaches the end of its life, its materials are kept within the economy wherever possible. These can be productively used again and again, thereby creating further value.

The circular economy expands on the traditional 3Rs approach (reduce, reuse and recycle) to consider a broader scope of actions, the multi-R approach: reduce, redesign, reuse, repair or renew, remanufacture, recycle, recover (energy and materials) and restore. All of which are attempted before sending the product to landfill (Fig. 1).

Given the pressure from many governments with increasingly stringent legislation and consumers' growing environmental concerns, the fashion industry is being forced to make changes towards more circular business models. Some brands only make small changes, which—in most cases—are similar to their competitors and are merely to comply with legislation. However, they promote and communicate their supposed interest in protecting the environment. Such actions are considered greenwashing, that is, public messages issued by a company that fabricates or exaggerates the product's ecological aspects or the organisation's commitment.

But other brands do adopt a real significant commitment to the environment and design their clothes based on eco-design principles, trying to promote slow fashion. Customers of such brands prefer to buy fewer clothes, appreciating the product's quality, paying a fair price and keeping the garments for longer.

2 The Ecological Footprint of the Clothes in Our Wardrobe

Negative impacts on the environment occur throughout a garment's life cycle:

- *Producing the raw materials.* Historically, cotton has been the most commonly used material. Although it is sourced naturally, its production has a major environmental impact because it requires a significant amount of land, water, fertilisers and pesticides. Currently, between 55 and 60% of clothes are made of plastic fibres, mainly polyester and nylon (Ellen MacArthur Foundation, 2017). These materials do not require farming land and use less water but they are obtained from petrol and have a high carbon footprint. Furthermore, these materials tend to shed microplastic fibres into wastewater that ends up in the seas and oceans.
- *Logistics to factories.* On many occasions, the production centres are located far away from the farming areas, requiring logistics and transport management with its resulting carbon footprint.
- *Manufacturing.* Major fashion brands subcontract their clothing manufacturing to factories located in countries with low labour costs. Materials are prepared, patterns are cut and sewn and dyed and the final product is washed and dried in these factories. This stage consumes a lot of energy and water, uses many chemicals (like dyes) and generates a lot of polluted water.
- *Logistics to points of sale.* From the factories, the clothes are taken to logistics centres and, from there, to the stores. During this stage, the garments are packaged and transported to countries and continents far away from the factories with its resulting carbon footprint.
- *Sales at the point of sale.* Sales must also be managed correctly to ensure efficient use of the energy and waste generated, mainly related to the product's packaging.
- *The customer's garment use.* As consumers, we also leave a considerable environmental footprint due to the water, energy and chemicals we use to wash, dry and iron the garment. A determining factor is how often we wash the garment.
- *Discarding the garment.* Depending on the consumer's behaviour, this is the stage when they decide if their item of clothing is to end its useful life in landfill or be reused or remanufactured through a circular economy pathway.

We shall now look at some data associated with these impacts. They were estimated by research conducted by public and private bodies, such as the European Environment Agency (EEA, 2019), McKinsey and BOF (2019) and McKinsey and GFA (2020):

- The fashion industry's value chain was responsible for 2.1 billion tonnes of greenhouse gases in 2018. It is an amount equivalent to the total emissions produced by the United Kingdom, Germany and France combined.
- According to several estimates, the fashion industry represents between 4% and 10% of the world's total greenhouse gas emissions.
- Between 15 and 35 tonnes of CO_2 equivalent are emitted for every tonne of textiles produced. From another perspective, this industry's emissions represent 654 kg of CO_2 equivalent per person per year.
- The textile industry uses 3500 chemicals, of which 750 are harmful and affect human health and 440 are hazardous for the environment. In that regard, between 10% and 20% of insecticides are used on crops farmed to produce textile raw materials.
- The textile industry is responsible for 20% of wastewater.
- It is also responsible for between 20% and 35% of all microplastics in our oceans.

To help us visualise it better, let us take a pair of jeans and a cotton t-shirt—wardrobe staples—as an example. According to several studies, manufacturing a pair of jeans outside of Europe requires 7500 litres of water, that is, approximately the same amount of water that an adult drinks over a period of ten years. What is more, 0.5 kilograms of chlorine and 10 kilograms of dye and other chemicals are used. And 2700 litres of water are needed for a t-shirt, mainly to farm the amount of cotton required.

3 Clothing That Gives Us Short-Lived "Happiness"

How and where clothes are manufactured is not the only important factor. We should also think about how much we buy. The fashion sector is one of the most representative of the mass consumption society in which we live. This sector often uses the "buy your happiness" strategy, giving way to the *fast fashion* phenomenon. Large multinationals try to produce large volumes of clothing at very low prices, even though the materials are extremely polluting and they do not guarantee, in many cases, adequate working conditions, and pay suppliers a fair price.

To sustain this production system, they must convince consumers to buy a lot and often. Before, brands would launch summer and winter collection campaigns, but new collections are being launched much more frequently. They work with the concept of planned obsolescence or, more specifically, with psychological obsolescence, that is, they try to make customers believe that the clothes they already own are no longer fashionable and they must renew part of their wardrobe even if that were to mean having more clothes than they need and discarding others that are still in a good condition.

Furthermore, there is the added pressure of commercial communication and price reduction strategies, which have become almost constant in most countries: summer sales, winter sales and mid-season sales, Black Friday and Cyber Monday, etc. (Fig. 2). In short, it is the opposite of the circular economy's R principle: reduce.

Fig. 2 Examples of adverts to encourage consumption. Source: own creation using pixabay.com

According to the Ellen MacArthur Foundation report (2017), the average person today buys 60% more clothing than in 2000. And not only do they buy more, they also discard more as a result. "Worldwide, clothing utilisation—the average number of times a garment is worn before it ceases to be used—has decreased by 36% compared to 15 years ago... In the last 15 years, clothing production has approximately doubled, driven by a growing middle-class population across the globe and increased per capita sales in mature economies".

As a result, the fashion industry is globally responsible for around 40 million tonnes of textile waste a year, most of which is either sent to landfill or incinerated (McKinsey & BOF, 2021). In the European Union, consumers discard about 11 kg of textiles per person per year and textiles recycling is low (EEA, 2019). It could be concluded that *the textiles system operates in an almost completely linear way: large amounts of non-renewable resources are extracted to produce clothes that are often used for only a short time, after which the materials are mostly sent to landfill or incinerated.*

4 Towards Sustainable Fashion: The Path Towards a Textile Circular Economy

In recent years, the textile and clothing sector has started to react to consumers' growing concerns and, more specifically, to pressure from tougher legislation. Some examples of the sector's collective initiatives are the Fashion Industry Charter for Climate Action (adopted during the 2018 United Nations Climate Change

Fig. 3 Examples of eco-certificates and eco-labels. Source: own creation from certifying organisations' websites

Conference where undersigning companies committed to reducing net-zero greenhouse gases by 30% by 2030) and the Fashion Pact (presented at the 2019 G7 Summit). In this context, major multinational brands are adopting circular economy actions, which we are going to split into four lines of action as follows:

(a) *Longer product life using lesser raw materials.*
The sector is working to develop technological processes that can manufacture clothes using fewer raw materials, dyes, water and energy (reduce). It is also working on alternatives to substitute traditional cotton and polyester for more sustainable materials and fibres (redesign). There are three redesign alternatives. On one hand, using organic cotton is almost 50% less intensive in polluting emissions than traditional cotton, due to the limited use of insecticides and fertilisers.

A second option is replacing synthetic fibres with forest fibres; the raw materials for such come from responsibly managed forests or species such as hemp, bamboo or cork. A third alternative is to use recycled materials. These could be recycled textiles (known as closed-loop recycling or garment-to-garment recycling) or by-products from other products unrelated to the sector, such as PET (polyethylene terephthalate) plastic from bottles (open-loop recycling).

Some organisations have created environmental certification systems or eco-labels that independently guarantee that clothes are totally or partially manufactured with these alternative materials (Fig. 3). However, it is still common to find the following messages on clothes labels: "we commit to minimize water and chemicals consumption in our denim pants" or "a washing process with new sustainable technologies has been used in these jeans reducing the consumption of water, chemicals and energy". These environmental statements from the manufacturer do not comply with the recommendations given in the international standard ISO 14021.[1] This standard acts as a voluntary code of best practice and recommends the following rules, among others:

- To avoid vague and imprecise statements.
- The three following requirements must be fulfilled for statements about recycled content: (1) state the percentage of recycled material, (2) separately state the percentage of the content recycled for the product and the packaging and (3) using a symbol is optional, but when used, it must be the Möbius loop.
- For statements about using fewer materials, energy or water, brands must indicate (a) the amount and reduction percentage, (b) the basis for comparison (sector average, another competitor, product's previous version) and (c) the product's life cycle where this reduction is achieved.

(b) *Lifetime extension of products and parts*. Before we consider a garment as waste, its useful life can be extended by performing any of the actions we will describe in more detail below. Some actions can be carried out by the consumers themselves and others by companies in the sector if they implement used clothing collection systems (recover):

- Reusing clothes for alternative and different uses.
- Repairing, reconditioning or renewing.
- Remanufacturing or reproducing.
- Redistributing. This concept includes all actions that allow the clothes to be used by different people. In many cases, redistribution is synonymous with the collaborative economy:
  - Buying and selling clothes in second-hand shops, which could be physical shops or on apps (like Vinted) or social media.
  - Renting clothes. In this case, fashion becomes a service, meaning that one item of clothing can be used by many users at different times. This alternative is ideal for outfits for weddings or special events or maternity wear and baby clothes.
  - Donating to individuals or charity shops.
  - Clothes swap.

(c) *Useful purpose of material at end of life*. If the item of clothing can no longer be used, reconditioned or remanufactured, it becomes textile waste, and alternatives must be found to recycle its fibres and components so that it can produce new clothes or other alternative textiles.

(d) *Compensation systems*. To comply with their climate change policies, some companies undertake actions to compensate for their greenhouse gas emissions. They try to restore part of the damage caused through actions like planting trees.

[1] https://www.iso.org/standard/66652.html. ISO 14021 is an international standard that acts as a self-regulatory code, guiding companies on how to write an ecological message in a commercial communication or advertising. This is to avoid greenwashing, that is, misleading or confusing messages for the consumer.

Fig. 4 Examples of sustainable and circular fashion. Source: own creation and pixabay.com

In recent years, major fashion multinationals have started adopting most of these four lines of action (Fig. 4). Some of the sector's retailers are introducing spaces in their stores to collect used clothes. For example, H&M was a pioneer in implementing this system in 2013, incentivising consumers to participate by offering them a discount on their next purchase. In 2019, the company also launched a suit-renting service in some of its stores.

After recovering the garments and classifying them in line with their conditions, manufacturers can launch different products to extend their useful life. For example, with the project *Move to Zero*, Nike collects own-brand items and gives consumers the option of donating them to a charity after a light clean, selling them as refurbished products after a slightly more complex repair and renewal process or recycling them to extract fibres and manufacture new products.

By recycling, fashion brands are launching clothing collections made using recycled materials, such as recycled textile fibres or recycled plastics. Brands are also introducing more collections including garments made with sustainable cotton or linen. Some companies have set goals that 100% or most of their garments will be sustainable by 2025 or 2030. However, sustainable collections currently represent only a small part of their businesses.

The brand Springfield (Tendam group) is an example of a company that is compensating greenhouse gas emissions. In 2020, it launched the *Bosque Springfield* initiative to restock forests and woodlands in Spain and Portugal. Every time a customer buys a garment associated with this initiative, Springfield plants a tree.

5 Ecoalf: A Sustainable Brand from the Start

Unlike major fashion brands that are starting to adopt eco-design and textile recovery measures (albeit to a small percentage of their collections), other smaller companies in the sector came to being with sustainability and circular economy at their core. One such company is Ecoalf.

Ecoalf was founded in Spain in 2009, making a name for itself with a slogan that represents its raison d'être: "Because There Is No Planet B". Its mission is to create recycled products with the same quality and design as the best non-recycled products on the market. It takes inspiration from "not being the biggest company in the world, but being the best company for the world". This was mirrored in the company being awarded the *B Corp Certification* in 2018. This certification recognises those companies that fight for a more inclusive and sustainable economy for all people and the planet, complying with demanding standards of social and environmental performance, transparency and legal commitment.

Ecoalf adopts the *slow fashion* philosophy by offering long-lasting products and refusing to continuously launch collections and offer discounts (Fig. 5). As the company website states, "sustainability, innovation, quality and timeless design is a must for Ecoalf as a sign of durability and responsibility. Fashion trends are not one of our core design principles. We believe good design is neither new or old. It is merely a bridge from the past to the future".

Fig. 5 Examples of Ecoalf advertising campaigns. Source: own creation using www.ecoalf.com

By applying eco-design to all of its products, Ecoalf invests large amounts in R&D&I and creates alliances and collaborations with other companies, ecological organisations, technology centres and public bodies. In this way, it has managed to develop technologies that manufacture clothing and footwear with recycled materials, receiving the *Global Recycled Standard* certificate for its use of the following:

- *Recycled PET*, mainly bottles. With this material, Ecoalf is reducing water consumption by 20%, energy use by 50% and CO_2 by 60% compared with conventional polyester. Some of the recycled PET that it uses comes from the project *Upcycling the Oceans*, developed by the Ecoalf Foundation. The objective is to clean oceans of marine debris with help from the fishing sector and give plastic waste a new life through recycling. In 2022, the project worked with around 3500 fishermen in 60 ports in four countries: Spain, Thailand, Greece and Italy. Once collected and taken to the port, the marine debris is classified and stored to later be recycled. PET plastics represent just part of this debris and are transformed into high-quality 100% recycled polyester, which Ecoalf calls "Ocean Yarn". The rest of the recovered debris is passed on to other waste management companies.
- *Recycled nylon.* Ecoalf avoids using virgin nylon by recovering and recycling discarded fishing nets, carpet remains and leftovers from pre-production nylon. With recycled nylon and using hydrogen bonds to seal in the microfibres, Ecoalf has developed a 0-Release yarn that practically does not shed microplastics when washed.
- *Recycled cotton.* Besides reducing waste, recycled cotton can also achieve the desired colour of a garment by simply mixing fibres of the same colour during the recycling process. In doing so, Ecoalf skips the dyeing process and reduces the amount of water used.
- *Recycled tyres.* Although the process to recycle tyres is very complicated, it produces very high-quality products, such as flip-flops.
- *Coffee grounds.* Coffee grounds are mixed with PET or recycled nylon so that the yarn naturally contains all the properties coffee can give—fast-drying, UV protection and odour control—without using chemicals.

Ecoalf has also started to open its own stores. In 2022, the company was present in several cities in Spain, as well as Berlin, Tokyo and Turin. Its green philosophy is also evident in how stores are designed: furniture is natural, employing recycled materials such as bamboo or recycled wood and iron from demolition works. Ecoalf has created green spaces inside its stores to improve air quality and employs efficient lighting systems. Furthermore, in its efforts to raise awareness, the stores have an *ACT Now* space, used for holding events and talks.

6 Conclusions

The fashion sector has been widely criticised in recent decades for infringing on labour rights and even human rights in the factories that are subcontracted to manufacture products in underdeveloped or developing countries. However, it was not until more recently that the general public became interested in its high environmental impact.

The unsustainable situation on Earth, especially due to climate change, is forcing current production and consumption models to change. There needs to be a shift towards sustainable fashion produced under circular economy principles. Changes in the sector will help make progress on several of the Sustainable Development Goals (SDGs) set by the UN for 2030. The sector mainly affects Goal 12, responsible consumption and production; Goal 13, climate action; Goal 14, life below water; and Goal 15, life on land.

The sector's major multinationals are already taking steps towards more sustainable fashion, but they often act due to legal pressure and only adapt part of their products and collections. What is more, they continue to employ *fast fashion* strategies to promote the mass consumption of clothing. On the contrary, small companies with 100% sustainable and circular business models are emerging, but they struggle to compete against the power of multinationals.

To a great extent, multinationals' progress towards sustainable fashion and the survival and success of small sustainable companies will essentially depend on market acceptance. Under which conditions are consumers willing to change towards sustainable fashion? Are you willing to change?

7 Discussion Questions

1. *Critical analysis of a fashion company's environmental information.* Choose a fashion brand and go to its website. Read its sustainability policy and identify the sustainable and circular fashion actions that it has implemented. Then, analyse if the information about these actions complies with the requirements set out in international standard ISO 14021[2] for self-declared environmental claims and if the company has any environmental certificates from external bodies. Do you think the company is doing good green marketing or is it just greenwashing?
2. *Debate: buying habits and clothing use.* To help you answer the questions at the end of the case, get into a group and discuss the following:
   (a) What are the most relevant criteria you consider when buying a garment? How much importance do you give to the price, brand or design? Do you also think about where the garment was manufactured or the materials used?
   (b) What importance do you give to sales periods and campaigns like Black Friday? Do you think you are influenced by psychological obsolescence?

[2] See ISO webpage: www.iso.org/standard/66652.html.

(c) How willing are you to buy clothes made of sustainable fabrics? What hurdles are stopping you from doing so?
(d) How willing are you to participate in the collaborative economy, that is, buying and selling second-hand clothes, donating or swapping clothes? What hurdles are stopping you from doing so?
(e) How willing are you to reduce the number of garments you buy?

Appendix: Teaching Notes for the Case

Synopsis

The fashion industry has a huge impact on the environment. According to the United Nations (UN), it is the second most polluting industry in the world. On one hand, a high volume of water and large quantities of chemicals are used during a garment's life cycle. What is more, the sector has a high carbon footprint given that factories are located in underdeveloped countries far from the main markets. But fashion is just the tip of the iceberg of the mass consumption society. The fashion industry's marketing strategies encourage impulse buying through psychological obsolescence and by creating a sense of well-being. We are talking about *fast fashion*—low-cost products with a short useful life.

A shift is needed in the fashion industry to include the circular economy's *multi-R approach*: reduce, redesign, reuse, repair or recondition, remanufacture, redistribute, recycle, recover and restore. This chapter explains how some major fashion brands are applying circular economy strategies, so we can reflect on how environmental sustainability has progressed. We will also consider if they are really eco-friendly marketing strategies or greenwashing.

In contrast to these fashion giants, we will analyse the case of the company Ecoalf. It is a company where eco-design is at the heart of all products and is a prime example of slow fashion. Under the slogan, "Because there is no planet B", Ecoalf manufactures clothes from fishing nets, plastics recovered from the sea and used tyres.

Potential Audience and Instructor's Material

The case has been developed for use in marketing management courses and in courses on corporate social responsibility and environmental management. It is appropriate for MBA and Executive Development Programs as well as advanced undergraduate courses. The case is appropriate for the courses that deal with the topics: "responsible marketing", "green marketing" and "fashion marketing".

Learning Objectives

1. To reflect on the environmental implications involved in companies' buying, manufacturing, marketing and location decisions applied to the fashion and textile industry in this case
2. To encourage debate about the mass consumption society and marketing practices that encourage such, as well as the consumers' readiness to change their buying and consumer habits
3. To understand the circular economy concept by comparing it to the current linear economy
4. To understand the different ways a circular business model can be designed through eco-design and materials and waste management

Time Frame for Class Discussion

This case can be taught in a session of between 60 and 90 minutes. Its duration depends on the number of students participating in the discussion activities, as well as the previous work that students have done before the face-to-face session.

Depending on the characteristics of the course, the teacher may consider the suggested activities to be carried out in the classroom or they may be proposed as activities to be carried out outside the classroom. In the same sense, the suggested activities can be carried out individually by each student or they can be carried out as team work.

Suggested Discussion Questions

1. *Critical analysis of a fashion company's environmental information.* Choose a fashion brand and go to its website. Read its sustainability policy and identify the sustainable and circular fashion actions that it has implemented. Then, analyse if the information about these actions complies with the requirements set out in international standard ISO 14021[3] for self-declared environmental claims and if the company has any environmental certificates from external bodies. Do you think the company is doing good green marketing or is it just greenwashing?
2. *Debate: buying habits and clothing use.* To help you answer the questions at the end of the case, get into a group and discuss the following:
   (a) What are the most relevant criteria you consider when buying a garment? How much importance do you give to the price, brand or design? Do you also think about where the garment was manufactured or the materials used?
   (b) What importance do you give to sales periods and campaigns like Black Friday? Do you think you are influenced by psychological obsolescence?

[3] See ISO webpage: www.iso.org/standard/66652.html.

(c) How willing are you to buy clothes made of sustainable fabrics? What hurdles are stopping you from doing so?
(d) How willing are you to participate in the collaborative economy, that is, buying and selling second-hand clothes, donating or swapping clothes? What hurdles are stopping you from doing so?
(e) How willing are you to reduce the number of garments you buy?

Analysis

Answer to Question 1. *Critical analysis of a fashion company's environmental information.*

This is an activity that students should do outside of classroom sessions. Ask each student (or small group of students) to choose a fashion brand or choose a group of brands that you would like the groups to analyse and give one to each student (or group).

Ask students to look for the sustainability section on the brand's website. After reading it, they will write a report that contains at least three sections:

- A short presentation about the brand: origin, geographical presence, types of products it sells, etc.
- A list and description of the sustainable and circular fashion actions that the brand has implemented, paying particular attention to the level of implementation.
- A personal reflection about whether the brand is really committed to sustainability or it is greenwashing. You should recommend students to consider if:
  - They are generalised measures in the sector.
  - The actions are to comply with a law or future law.
  - The actions are communicated in an effective, believable way according to the code set out in the standard ISO 14021 for self-declared environmental claims (https://www.iso.org/standard/66652.html).
  - The brand has any eco-certificates or labels awarded by an independent body.

You must decide how to conclude the activity: you can ask the students to hand in their reports or you can organise a session in which each student gives a presentation of between 5 and 10 minutes on the conclusions from their research.

Answer to Question 2. *Debate: clothing buying and use habits.*

The objective of this activity is to find out how we buy clothes as consumers, how we are influenced by marketing strategies and companies' psychological obsolescence and under which conditions we would be willing to buy sustainable fashion and even reduce how much fashion we consume.

This is an activity to encourage students to reflect, analyse and debate. This activity can be carried out in two ways:

(A). *In-class groups.* The activity can be carried out as follows:
- Invite the class to form groups or put the class into groups and give them a set time to discuss the topics (approximately 10–15 minutes). Groups of 4–6 are recommended.
- After 10–15 minutes, give the group another 3–5 minutes so they can summarise the opinions discussed. One of the students from each group will act as a spokesperson and will note down the summary.
- Then, all the students will come back together and each spokesperson will present their group's ideas.
- Conclude the activity by summarising the main ideas and opinions.

(B). *Individual project.* Outside of the classroom, each student must ask 4–5 people their opinions about the subject matter, preparing a series of open-ended questions. Each student must hand in a report including the following:
- A brief description of the group: how many people, ages, sex, etc.
- A summary of the main ideas and opinions: recommend students to include direct quotes from the participants

References

EEA. (2019). *ETC/WMGE Report 6/2019: Textiles and the environment in a circular economy.* European Environment Agency. https://www.eionet.europa.eu/etcs/etc-wmge/products/etc-wmge-reports/textiles-and-the-environment-in-a-circular-economy

Ellen MacArthur Foundation. (2017). *A new textiles economy: Redesigning fashion's future.* https://ellenmacarthurfoundation.org/a-new-textiles-economy.

Ellen MacArthur Foundation. (2022). https://ellenmacarthurfoundation.org.

McKinsey & BOF. (2019). The State of Fashion 2020. Mckinsey and The Business of Fashion. www.mckinsey.com/~/media/mckinsey/industries/retail/our%20insights/the%20state%20of%20fashion%202020%20navigating%20uncertainty/the-state-of-fashion-2020-final.pdf.

Mckinsey & BOF. (2021). *State of Fashion 2022: An uneven recovery and new frontiers.* Mckinsey and The Business of Fashion. https://www.mckinsey.com/industries/retail/our-insights/state-of-fashion.

McKinsey & GFA. (2020). How the fashion industry can urgently act to reduce its greenhouse gas emissions. McKinsey and Global Fashion Agenda.

Sustainable Business Models in Fashion Industry: An Argentine Social Enterprise Fostering an Inclusive and Regenerative Value Chain

María Fernanda Figueroa, Adriana Marina,
and Edison Irving Benites Leiva

Learning Objectives

- Learn about new organizational forms tackling sustainable development challenges in a more holistic way, analyzing the systemic change model of Animaná and HechoxNosotros.
- Delve into circular economy, exploring the regenerative production processes of Animaná and HechoxNosotros.
- Dig deeper into the sufficiency economy, analyzing Animaná's communication strategies on consumption sufficiency.
- Understand how sustainable business models work, exploring Animaná and HechoxNosotros' strategies to create, deliver, and capture economic, social, and environmental value simultaneously.
- Visualize the link between a social-environmental problem and a business value proposition incorporating such problems into their core business, not as a peripheral activity, but as a central one.

1 Introduction

The impact of human activity on natural ecosystems has exceeded several planetary boundaries, including biodiversity and climate change (Steffen et al., 2015). In addition, emerging countries are still struggling to achieve social and economic development, facing problems such as poverty and inequality, which the pandemic

M. F. Figueroa (✉)
Facultad de Ciencias Empresariales, Universidad Austral / CONICET, Buenos Aires, Argentina
e-mail: mfigueroa@austral.edu.ar

A. Marina
Founder of Animaná and HechoxNosotros, Buenos Aires, Argentina

E. I. Benites Leiva
Partner of HechoxNosotros, Lima, Peru

E. C. Bianchi et al. (eds.), *Responsible Consumption and Sustainability*, Springer Business Cases, https://doi.org/10.1007/978-3-031-30742-3_4

COVID-19 has exacerbated. In consequence, there is a claim to change to sustainable production and consumption systems, staying within the planetary boundaries while meeting human needs (Raworth, 2017).

The fashion industry is not alien to these problems, with environmental pollution and labor exploitation deeply rooted in their practices. It is considered as the second most polluting industry in the world (United Nations, 2022) and is deemed responsible for several human rights violations due to lack of integration and control in its value chain (Vásquez Martínez, 2021).

In addition, the pandemic has fueled new trends in sustainable fashion consumption demanding a level of transparency and commitment that "businesses as usual" are not prepared to satisfy (McKinsey & Company, 2021). The traditional approach of corporate social responsibility, implementing peripheral social and environmental actions, neither can deal with the magnitude of sustainable development challenges.

Therefore, sustainable business models are needed, with sustainability playing a strategic role in the core business and the use of the market force to solve social and environmental problems (Baldassarre et al., 2017).

Based on sustainable business model research, and through a case study of the Argentine social enterprise Animaná and their associate NGO HechoxNosotros (HXN), [1] this work aims at analyzing how a social enterprise in the fashion industry reaches international sustainable luxury markets while generating a sustainable value chain, and which their strategies are to create, deliver, and capture economic, social, and environmental values simultaneously.

This case study will contribute to literature on sustainable business models, analyzing the strategies to create, deliver, and capture value of a specific sustainable business, called social enterprise (Doherty et al., 2014).

In addition, it will extend the literature on sufficiency economy (Niessen & Bocken, 2021) and circular economy (Ellen Macarthur Foundation, 2015), delving into marketing and production strategies that promote reduction of consumption and aim to keep products, components, and materials at their highest utility and value at all times, decoupling global economic development from finite resource consumption, to mitigate associated environmental degradation, in line with planetary boundaries, while maintaining solid social and well-being standards for all citizens on an equal basis.

[1] Animaná means "a place in heaven" in Kakane, a derivative of the Quechua dialect spoken by natives from Northern Argentina.

2 Case Development

2.1 The Fashion Industry

The fashion industry generates USD3 billion on an annual basis, accounting for the 2% of the global GDP and employing more than 300 million people (Ellen Macarthur Foundation & Circular Fibres Iniciative, 2017). However, 85% of workers are not covered by any formal protection, and they work under unfavorable conditions (ILO, 2015).

On the other hand, fashion is the second most polluting industry in the world responsible for 10% of the global carbon emissions and the 31% of the microplastic ocean pollution (Niinimäki et al., 2020). In addition, the 87% of the material used for clothing manufacturing is burned or disposed of in landfills, resulting in several and negative environmental impacts (Ellen Macarthur Foundation & Circular Fibres Iniciative, 2017).

Faced with these problems, in 2009, Adriana Marina (economist and entrepreneur) founded Animaná, [2] an Argentine social enterprise certified as a B corporation, aiming at connecting local artisans and international consumers. [3, 4]

2.2 Animaná: Inclusive and Sustainable Business Model

Animaná is a sustainable luxury brand producing natural fiber garments from Patagonia and the Andes, employing 8500 artisans, 50 small and medium enterprises, and 15,000 natural fiber producers throughout its value chain, implementing circular economy production strategies and fostering consumption sufficiency with timeless collections reducing seasonality in the fashion system. Currently, Animaná has a boutique store in Buenos Aires and exports to France, the United States, and Canada, with ten full-time collaborators.

Adriana (see Image 1) lived her childhood surrounded by natural landscapes and local wildlife from Patagonia, in Santa Cruz, Argentina. But, at the same time, she witnessed the social inequality in the treatment of rural communities and the loss of Andean arts due to the imminent exodus toward the cities. Therefore, Adriana started this project together with artisan communities, cooperatives, and Andean knitters, in the search of rescuing and reappraising natural fibers and ancient techniques.

Together with Animaná, Adriana also founded Hecho por Nosotros (HXN), [5] a nonprofit NGO with consultative status at the United Nations (United Nations, 2016)

---

[2] https://animanaonline.com.ar/

[3] https://fashionunited.com.ar/noticias/empresas/a-traves-de-creative-experience-tres-marcas-argentinas-se-presentan-en-le-new-black/2022030732615

[4] https://www.comex.icbc.com.ar/categoria/premio-comex

[5] https://es.hechoxnosotros.org/

Image 1 Adriana Marina wearing Animaná clothes. Source: Animaná and Hecho por Nosotros (2021)

focused on establishing fashion as a means of change and local development, as well as on educating and promoting the principles of sustainable fashion.

In addition, HXN is focused on connecting the global professional network to grassroot organizations, working as a facilitator, and supporting the co-creation of theoretical tools which will contribute to the paradigm change of sustainable fashion and will prepare communities to integrate them into the value chain of global fashion. Currently, HXN has 5 full-time collaborators and more than 500 volunteers. Animaná and HXN have as common purpose reappraising the value chain of natural fibers to grow development of small local producers and work for an environmentally sustainable fashion industry. HXN facilitates the development of Andean small producers' capabilities and co-creates with them a value proposition by giving advice on global market demands. Animaná is the bridge between producers and consumers and deals with the production commercialization.

Animaná is an Argentine social enterprise (SE) [6] certified as a B corporation, [7] which proves its engagement with triple impact (social, environmental, and economic).

Its commercial activity allows them to maintain their social mission, which is reappraising the Andean culture, by offering development opportunities to local producers, and to maintain their environmental mission, which is reducing the

[6] Social enterprises are organizations which find a solution to social and/or environmental problems through the commercialization of goods and services in the market. Social and/or environmental problem-solving is the core business and not a peripheral activity. This is the main difference with corporate social responsibility.

[7] https://www.bcorporation.net/en-us/find-a-b-corp/company/animana/

pollution generated by textile manufacturing with natural fibers and dyes, encouraging responsible consumption as well. Animaná is a special type of SE since it integrates a hybrid organizational group together with the NGO, HXN. Both have a common purpose, but it is accomplished through different paths.

> To accomplish this mission, implying a systemic change, I needed to create a virtuous ecosystem, a place bringing together all stakeholders to integrate them into the whole value chain and make it visible. So, I needed to create a space open to speak with institutions, such as the United Nations and universities, bringing together several professionals around the world who had this same dream but found it difficult to articulate. I needed to create a full facilitation platform and that is HXN.
>
> On the other hand, I have never wanted to be just a sustainability advocate, rather I wanted to make this transformation effective in the reality of artisans, producers and the local community, and that is Animaná. Therefore, Animaná and HXN together allow me to carry forward my mission.
>
> (Adriana Marina, founder of Animaná and HXN)

Animaná develops different strategies of circular economy [8] helping to reduce pollution in productive processes such as the care of shearing processes to preserve biodiversity, the use of organic fibers, the utilization of natural colors from fibers or the use of plants for clothes dyeing, and the utilization of craft techniques and instructions of use for making clothes durable.

However, the present case is focused on the consumer perspective, analyzing those strategies implemented by Animaná which encourage consumption sufficiency. Animaná targets a segment which not only appreciates the design quality but also the manufacturing methods. It is about a more engaged and responsible consumer in their purchases.

Adriana is convinced of the huge opportunity existing within the simple luxury market evoking the luxury of giving access to the origins, to nature, which has been lost in this fast, automatic, and cost-reduction culture, where now natural products are considered as luxurious items. These are natural fibers, products made with wisdom, far from time pressure, respecting the culture and honoring the land.

In addition, the business model of Animaná aims at keeping or reducing the consumption curve leading to environmental degradation, while social and well-being needs are satisfied in an equitable manner (Niessen & Bocken, 2021). While sufficiency has an inherent purpose transforming the consumer behavior to consume less and better, it is interesting to analyze how companies encourage such sufficiency. Therefore, Animana's business model will be analyzed within the "Business Framework for Sufficiency" developed by Niessen and Bocken (2021), as shown below (Table 1).

Columns of Table 1 represent three messages from the company to the consumer:

---

[8] A circular economy is an economic model which from design aims to keep products, components, and materials at their highest utility and value at all times, distinguishing between technical and biological cycles, ultimately decoupling global economic development from finite resource consumption (Ellen Macarthur Foundation, 2015).

Table 1 Business framework for sufficiency

	Rethink *Consume differently*	**Reduce** *Consume less*	**Refuse** *Do not (over)consume*
Less clutter *Simplified and less*	**Green alternative**		
Less speed *Slower and more reliable*		**Timeless clothing-better quality textiles**	
Less distance *Regional and disentangled*	**Cultural alternative**		
Less market *Beyond commerce*			

Source: Own elaboration based on the Niessen and Bocken's (2021) model adapted to Animana's strategies

- *Rethink* consumption invites to consume more sustainably but not necessarily to reduce consumption.
- *Reduce* consumption invites to consume fewer products.
- The most radical message is *Refuse* consumption which implies not consuming at all.

Rows indicate four sufficiency strategies which may be implemented by companies:

- *Less clutter* refers to designing simpler products with less materials or pieces, and even offering the service of product use instead of product ownership, by renting, leasing, or sharing products.
- *Less speed* means more durable products which will not be replaced by new products in a short time.
- *Less distance* puts emphasis on decentralized value chains, appreciating local products.
- *Less market* implies going beyond commerce, encouraging consumers to produce on their own what they need or generate collaborative consumption circuits.

Animana's communication strategies within its segment are placed in the following model quadrants:

- *Green Alternative*
  Animaná encourages consumers to *Rethink* their fashion consumption habits, fostering the consumption change from clothes made of non natural materials to clothes made of natural materials and from standard and global clothes to clothes full of history and local culture. Animaná remarks the benefits of natural fibers and dyes for the own consumer's health, favoring skin care, for not using synthetic fabrics, and planet care, since camelid fibers, wood, cotton, and linen

Image 2 Animaná products. Source: Animaná and Hecho por Nosotros (2020)

are renewable resources, and natural dyes reduce the chemical pollution generated by the industry. Finally, since clothes are made of natural dyes and fibers, [9] they are biodegradable, simplifying the product end-of-life process.

- *Cultural Alternative*

  Animaná highlights the positive features of wearing culture clothes full of Patagonic and Andean history. It also remarks the positive impact of each purchase on the development of local community capabilities which, through a model with fewer intermediaries, help small producers to receive a fair price for their work.
- *Timeless Clothing-Better Quality textiles*

  Animaná invites to *Reduce* consumption, framed within slow fashion, [10] inviting to consume timeless clothing, not tied to such famous fast-fashion seasons. This helps consumers to get rid of such needs of being fashionable leading to a wardrobe renewal every year.

  On the contrary, Aninama's timeless designs are thought to last through the years offering comfort and well-being, since by using natural fibers and ancient manufacturing techniques, they make better quality and more durable clothes (see Image 2).

[9] https://animanaonline.com.ar/pages/nuestras-fibras

[10] https://www.slowfashionnext.com/blog/predicciones-de-moda-sostenible-para-el-2020/

2.3 The Sustainable Value Chain

2.3.1 Animaná

Animaná has as its main objective building a sustainable project, going hand in hand with ancient knowledge from Latin American artisans and textile producers, prioritizing sustainable production processes, while generating an ethical, sustainable, and transparent value chain. To achieve this objective, Animaná deploys a series of specific actions:

Reappraisal of ancient knowledge and work with local communities: Throughout modern history, societies have been developed from the Western blinkers, making invisible and disqualifying all such ancient and local knowledge alien to the Western structure of thought. However, currently, these practices and forms of thought become effective and allow thinking in other types of society and forms of production. In this respect, Animaná seeks to uplift all that ancient and local wisdom from artisans and textile producers preserving ancient weaving techniques and skills, with unique processes resulting in diverse textures and yarns, and that through a co-creative work, they may build a contemporary proposal that becomes attractive to the global market, while maintaining their traditions, authenticity, designs, and natural colors. Since 2009, Animaná, together with HXN, has been bringing together cooperatives, artisans, and producers, creating alliances, and fostering a direct collaboration with each of the local communities, forming part of an 8500-artisan network. This holistic approach with no intermediaries allowed local communities to triplicate their production capacity, generating higher profits and turning some artisan families into local entrepreneurs.

Sustainability from natural fibers and production traditions of local communities: Natural fibers from camelids in the Andes form part of the group of the finest fibers in the market, well known by their quality in the fashion industry, and represent an alternative to synthetic fibers.

To take care of natural fibers, Animaná closely supervises the camelids handling (see Image 3), checking that producers preserve their species and natural environment. The camelid shearing is completely manual, avoiding the use of any kind of environmentally polluting machinery.

On the other hand, the use of chemicals harmful for dyeing is avoided by implementing the Andean dyeing tradition with local flowers and plants; [11] in addition, its proposal is focused on natural color collection, going from natural black to different gray and brown tones.

Clothing manufacturing in Animaná is mainly manual, using different handlooms and tools made by their own communities using their ancient techniques (see Image 4).

---

[11] They use flowers and plants like kinsakucho, molle, eucalyptus, and chilca and some insects like cochineal.

Image 3 Vicuñas in their natural habitat. Source: Animaná and Hecho por Nosotros (2020)

In this way, Animaná respects traditions from local communities, using natural and organic yarns and dyes which avoid chemical waste in rivers, desertification caused by camelids' exploitation, and imbalance in biodiversity.

Encourage responsible consumption: From its beginning, Animaná seeks different means for disclosing its production processes while inviting consumers to become more curious about what they consume. Until 2020, Animaná and HXN hosted more than 60 events, workshops, and webinars with more than 500 academics and experts in sustainable fashion, reaching almost 100,000 people. [12]

In every purchase, Animaná intends that consumers become aware of the clothing features: origin, composition, yarn, dye, and manufacturing. This point is sought to be powered by the technology project for traceability and transparency led by HXN and explained in the following section. In addition, its focus on timeless designs fosters responsible long-lasting consumption, offering comfort and well-being to consumers. Through such actions as *#animanacare* and *#animanalonglife*, the company provides care instructions for clothing and for making them more durable, and if clothing is damaged, they offer to repair them.

[12] https://es.hechoxnosotros.org/_files/ugd/0fbe27_64fad1e65cf74c8abd0a038ccf960bc1.pdf

Image 4 Distaff spinning technique. Source: Animaná and Hecho por Nosotros (2020)

Commercialization: Animaná opened two boutique stores, one located in Palermo (Buenos Aires, Argentina) and another in Paris (France). It also exports to France, the United States, and Canada by participating in international fairs. [13]

Clothing prices range from USD 64 to 200; house linen, such as blankets, range from USD 200 to 500; and other accessories vary between USD 20 and USD 80. [14]

2.3.2 HechoxNosotros

The main activity of HXN consists in preparing local communities to form part of the global market and the fashion value chain; for that, it conducts a series of activities supported in an ecosystem where Animaná is part, but HXN is the main responsible.

Development of capabilities and co-creation: HXN contributes to the training and development of the necessary skills for the adequate integration of artisans, producers, and micro-, small-, and medium-sized enterprises (MSMEs) to the fashion value chain. For that, HXN has a network of more than 500 volunteer professionals from more than 34 countries in the world who make available their

[13] https://www.lanacion.com.ar/economia/animana-revalorizar-raices-exportar-sustentabilidad-nid2287530/

[14] Dollar official exchange rate as of June 1, 2022: 125.25 Argentine pesos

knowledge and experience to local communities; they have been offering 150,000 working hours so far since their beginnings until 2020. [15]

First, professionals make a diagnosis of the needs of small producers' communities, and then, they develop training plans adapted to each community. They work on topics related to fashion, textiles, accounting, administration, marketing, market research, social media, and branding, among others. Working with artisans and textile producers also implies a whole learning for the volunteer professionals' network, who seeks to co-create solutions while respecting and reappraising small producers' perspective and knowledge.

Technology for traceability and transparency: HXN has been developing since some years ago a project using the blockchain technology to certify the clothing traceability and to give consumers information about the place of origin and distribution of products, from their beginning to their final stage. This project is not intended to benefit only Animaná, but it also represents huge possibilities for organizations lacking integration in the supply chains, providing them more visibility while measuring and communicating in a tangible manner the product traceability.

Financing: HXN, in alliance with financial institutions such as IOV Lab [16] and Quipu, [17] seeks to provide financing for the working capital of such communities participating and forming part of the Animana's network of artisans and textile producers.

Research: HXN drives different research projects to achieve a deep knowledge of the social, economic, and environmental problems related to the textile industry, since it is essential for the development of small producers' capabilities.

2.4 Conclusions

Changing the system means altering in a permanent manner such perceptions, conducts, and structures that used to cause problems in the past. In social innovation, systemic change is commonly called "Impact," referring to the results achieved at a community or social level which are long-lasting and generate a significant change because they attack the deepest causes of a problem. In the present case, the impact sought is to improve the social and economic development of small producers living in the Andes and Patagonia and to reduce the environmental pollution caused by the fashion industry.

Both problems (social and environmental) are complex since they have multiple and global causes; therefore, they cannot be solved only by Animaná or HXN in the short term; instead, they require the effort of several stakeholders tackling the problem from different sides and perspectives for many years, at a local and global

[15] https://es.hechoxnosotros.org/_files/ugd/0fbe27_64fad1e65cf74c8abd0a038ccf960bc1.pdf

[16] https://www.iovlabs.org/

[17] https://soyquipumarket.com/

level. So small social and environmental changes generated by the joint action of several actors are turned into a systemic change. Thirteen years ago, Adriana Marina started the path of facing the problem of the poor working conditions suffered by fashion industry's workers in low-income countries and the problem of pollution generated by such industry at a global level. From the beginning, she knew that they were too big and complex problems to address from her own organization or from the social or business logic separately.

Therefore, through the foundation of the social enterprise Animaná and the NGO, HXN, she created a space for discussion, gathering, and transformation to convert fashion in a means of local development, working hard to build a chain of small producers who had the necessary capabilities to provide global markets in exchange of fair prices, maintaining their culture and ancient planet-caring techniques, while educating and disclosing the principles of sustainable fashion in Latin America and around the world. Although there are many satisfactory results (the growth of the artisans' network and improvements in the social and economic conditions, the increase of stakeholders involved in the HXN network, and the growing trend of conscious consumers representing huge market opportunities), the path has not been easy, and there are still many challenges to face.

For Animaná, reaching more consumers and fundraising pose even more difficulties, with many points to improve, such as the development of online marketing strategies to grow local and internationally, and to form alliances with companies to sell corporate gifts.

From HXN, two problems are viewed. Firstly, the lack of an economically sustainable model hindered by the fact that impact investors remain searching for results from the logic of a nonsustainable fashion system (current system), resulting in many NGOs going greenwashing and giving up the goal of a systematic change. Secondly, the lack of marketing resources allows to gain more presence and reach more key stakeholders to develop a varied and engaged ecosystem for a real systemic change.

3 Discussion Questions

1. Which is the value proposition of Animaná and HXN, and in which way are they related so that the social-environmental problem solution is within the core business?
2. Which sustainable development goals do the referred value propositions contribute to?
3. Analyzing the main business strategies of Animaná, how business success is compatible with the consumption sufficiency strategy?
4. Which actions could Animaná develop to address business and financing challenges currently facing while maintaining its sustainable essence?
5. Do you think that the systemic change model of Animaná and HXN is applicable to other industries which you are familiar with? Which challenges and opportunities do you detect?

Appendix: Teaching Notes for the Case

Synopsis

This case contributes to sustainable business model study, analyzing how an Argentine social enterprise (Animaná) and NGO (HechoxNosotros) in the fashion industry reach international sustainable luxury markets while generating a sustainable value chain and delivering into their strategies to create, deliver, and capture economic, social, and environmental values simultaneously.

In addition, this case study extends to literature on sufficiency and circular economy, digging into marketing and production strategies that promote consumption reduction and ensure the highest utility and value of products, components, and materials, always decoupling economic development from finite resource consumption.

Potential Audience and Instructor's Material

This case has been developed for use in the framework of management courses, in general and in sustainable business management or corporate social responsibility (CSR) seminars, in particular. The potential audience for this chapter is twofold:

- Undergraduate students pursuing degrees in business administration
- Graduate students pursuing MBA and masters, executive programs, or seminars in sustainable business or corporate social responsibility (CSR).

Learning Objectives

- Learn about new organizational forms tackling the sustainable development challenges in a more holistic way, analyzing the systemic change model of Animaná and HechoxNosotros.
- Delve into circular economy, exploring the regenerative production processes of Animaná and HechoxNosotros.
- Dig deeper into the sufficiency economy, analyzing Animaná's communication strategies on consumption sufficiency.
- Understand how sustainable business models work, exploring Animaná's and HechoxNosotros strategies to create, deliver, and capture economic, social, and environmental value simultaneously.
- Visualize the link between a social-environmental problem and a business value proposition incorporating such problems into their core business, not as a peripheral activity but as a central one.

Time Frame for Class Discussion

This case can be taught in one session of 60–100 minutes, depending on the number of students participating in the discussion activities, as well as the previous work that students have done before class.

Suggested Discussion Questions

1. Which is the value proposition of Animaná and HXN, and in which way are they related so that the social-environmental problem solution is within the core business?
2. Which sustainable development goals do the referred value propositions contribute to?
3. Analyzing the main business strategies of Animaná, how business success is compatible with the consumption sufficiency strategy?
4. Which actions could Animaná develop to address business and financing challenges currently facing while maintaining its sustainable essence?
5. Do you think that the systemic change model of Animaná and HXN is applicable to other industries which you are familiar with? Which challenges and opportunities do you detect?

Analysis

Answer to question 1. *Which is the value proposition of Animaná and HXN, and in which way are they related so that the social-environmental problem solution is within the core business?*

Ideas to work on this question:

Corporate social responsibility is commonly associated with a business logic placed in the organization's core and the social mission in the periphery. Nonprofit organizations are associated with the social mission placed in the core and the business logic in the periphery. Finally, social enterprises, for being hybrid organizations, put both logics in the strategic core.

Animaná and HXN start from the recognition of an industry with a fragmented value chain, where producers of natural fibers, artisans, and MSMEs (root and central basis of this industry) are in vulnerable and poor situations leading to the systematic loss of their holistic and biodiversity production and consumption models.

Therefore, the value proposition of Animaná and HXN is focused on integrating the fashion value chain, attempting, through technology, to maximize artisans, producers, and MSMEs' knowledge, often marginalized in the global fashion economy, validating their wisdom, sustainable techniques and processes, and natural fibers. In this manner, Animaná and HXN contribute to an inclusive model based on sustainability, circularity, and co-creation. This approach integrates the social mission to the business mission, both in the core of the organization's strategy.

Answer to question 2. *Which sustainable development goals do the referred value propositions contribute to?*

Material to work on this question: https://www.un.org/sustainabledevelopment/es/objetivos-de-desarrollo-sostenible/

Although most of the sustainable businesses address several goals at the same time, the most remarkable sustainable development goals are mentioned in this study case.

Goals 4 and 5: Quality Education and Gender Equality

HXN educates artisan women who had no access to formal education, so they are often marginalized and have low remunerations and little participation at their workplaces. Access to education spaces facilitated by HXN allows them to acquire the necessary tools to improve their social and economic situation.

Goal 8: Decent Work and Economic Growth

The value proposition of Animaná and HXN empowers small artisans, so they have a leading role in their economic and social development, helping them not only to improve their individual enterprises but also to improve the conditions of the community where they live.

Goal 11: Sustainable Cities and Communities

They promote the harmony of artisans and nature living together, without using chemicals, maintaining a healthy soil, avoiding diseases, and helping artisans to remain more time working on their home communities and avoiding their migration to cities in search of better opportunities.

Goal 12: Responsible Consumption and Production

From the production side, HXN trains small producers in regenerative and ecological practices, in addition to their own ancient traditions already contemplating the care for nature. Animaná also helps that these production capabilities consider such demands from luxury markets.

Both organizations work together for the consumers' awareness, so that they are more informed on the origins of products they buy, and they are more engaged with the development of people who manufacture such products.

Finally, since they are in the slow fashion framework, they discourage consumption by promoting less consumption or sufficiency in the fashion consumption.

Goals 13: Climate Action

By implementing production practices of a circular economy, Animaná and HXN contribute to the carbon dioxide reduction in several manners, mainly by reducing soil and ocean pollution with the use of natural fibers and dyes and by discouraging consumption.

Goal 17: Partnerships for the Goals

HXN is an emblem of collaborative work for systemic change; it is a platform bringing together several stakeholders to change the current fashion paradigm such as civil society organizations; international organizations; universities; micro, medium, small, and large enterprises; governments; and citizens.

Answer to question 3. *Analyzing the main business strategies of Animaná, how business success is compatible with the consumption sufficiency strategy?*

This is a question with an open answer. The purpose is that students analyze the consumption sufficiency as a business strategy and that they rethink the prevailing logics of business success.

Animaná's proposal considers business success as the development of a fair and regenerative value chain. Accordingly, its business strategy targets the luxury market comprised by people of a high social and economic level and who value simple, ethical, and environmentally friendly items and who are willing to pay a higher price for products offering such value. There is the danger of greenwashing, a communication strategy which seeks to emulate these new practices with the aim of showing consumers that they are respectful of the environment and its social causes, while maintaining a fragmented, unfair, and polluting value chain.

Answer to question 4. *Which actions could Animaná develop to address business and financing challenges currently facing while maintaining its sustainable essence?*

It is a question with open answer which intends that students think about solutions in the following two points, trying that such solutions find a balance between social, environmental, and economic logics:

1. Improve communication and marketing strategy:
   (a) In Animaná: to reach more consumers in the luxury market
   (b) In HXN: to reach key stakeholders which allow creating a varied ecosystem engaged with systemic change
2. Identify new forms of financing for triple impact organizations.

Answer to question 5. *Do you think that the systemic change model of Animaná and HXN is applicable to other industries which you are familiar with? Which challenges and opportunities do you detect?*

This is a question with an open answer. The purpose is that students analyze the implications of applying this model to other industries which they are familiar with (for work, profession, volunteering, etc.).

The food industries, such as agriculture and livestock, are intrinsically joined to the work done by Animaná and HXN with the communities, being an essential part for their integral development. However, it is important that systemic change starts from their own communities; it should not be a top-down imposition.

This implies rethinking our education system and the companies, producers, and consumers' role in the system, which are still based on centralized logics of power and knowledge and on the creation of "false dichotomies."

References

Baldassarre, B., Calabretta, G., Bocken, N., & Jaskiewicz, T. (2017). Bridging sustainable business model innovation and user-driven innovation: A process for sustainable value proposition design. *Journal of Cleaner Production, 147*, 175–186. https://doi.org/10.1016/j.jclepro.2017.01.081

Doherty, B., Haugh, H., & Lyon, F. (2014). Social enterprises as hybrid organizations: A review and research agenda. *International Journal of Management Reviews, 16*(4), 417–436. https://doi.org/10.1111/ijmr.12028

Ellen Macarthur Foundation. (2015). *Towards a circular economy: economic motives for an accelerated transition.* https://ellenmacarthurfoundation.org/towards-a-circular-economy-business-rationale-for-an-accelerated-transition

Ellen Macarthur Foundation., & Circular Fibres Iniciative. (2017). *A new textiles economy. Redesigning fashion´s future.* https://emf.thirdlight.com/link/2axvc7eob8zx-za4ule/@/preview/1?o

ILO. (2015). *Protecting workers in a changing world of work.* https://www.ilo.org/wcmsp5/groups/public/---ed_norm/---relconf/documents/meetingdocument/wcms_358293.pdf

McKinsey & Company. (2021). *Consumer sentiment on sustainability in fashion.* https://www.mckinsey.com/industries/retail/our-insights/survey-consumer-sentiment-on-sustainability-in-fashion

Niessen, L., & Bocken, N. M. P. (2021). How can businesses drive sufficiency? The business for sufficiency framework. *Sustainable Production and Consumption, 28*, 1090–1103. https://doi.org/10.1016/j.spc.2021.07.030

Niinimäki, K., Peters, G., Dahlbo, H., Perry, P., Rissanen, T., & Gwilt, A. (2020). The environmental price of fast fashion. *Nature Reviews Earth & Environment, 1*, 189–200. https://doi.org/10.1038/s43017-020-0039-9

Raworth, K. (2017). *Doughnut economics: Seven ways to think like a 21st-century economist.* Green Publishing.

Steffen, W., Richardson, K., Rockström, J., Cornell, S. E., Fetzer, I., Bennett, E. M., Biggs, R., Carpenter, S. R., De Vries, W., De Wit, C. A., Folke, C., Gerten, D., Heinke, J., Mace, G. M., Persson, L. M., Ramanathan, V., Reyers, B., & Sörlin, S. (2015). Planetary boundaries: Guiding human development on a changing planet. *Science, 347*(6223), 736. https://doi.org/10.1126/science.1259855

United Nations. (2016). *ECOSOC designation of consultative status to Hecho por Nosotros.* https://esango.un.org/civilsociety/showProfileDetail.do?method=showProfileDetails&tab=1&profileCode=631203

United Nations. (2022). *UN alliance for sustainable fashion.* https://unfashionalliance.org/

Vásquez Martínez, L. (2021). Stained supply chains: Labor exploitation in the textile industry of China and India. Online Journal Mundo Asia Pacifico, 10(19), 85–97. file:///C:/Users/Acer/Downloads/Cadenas+de+suministro+manchadas.pdf.

Part II

Environment

Want to Be Rewarded for Recycling? With RECICLOS You Can!

M. Manuela Palacios-González and M. Soledad Janita-Muñoz

Learning Objectives
The general objective of this paper is to raise the profile of the RECICLOS initiative devised by the non-profit organisation responsible for coordinating light household packaging waste recycling in Spain, Ecoembes, through its laboratory *TheCircularLab*.

It also seeks to achieve the following more specific objectives:

- Reflect on the change that is currently taking place in consumption habits, giving rise to "socially responsible consumption" and its links with waste packaging recycling, as well as understanding the importance of this type of consumption in the future strategies of international and European organisations.
- Analyse waste recycling data in the European Union and Spain.
- Learn more about the non-profit organisation Ecoembes, responsible for coordinating light household packaging waste recycling in Spain, its mission, pillars and links with other similar organisations across Europe.
- Study the RECICLOS initiative, its operation, main achievements and the perceptions of the various interest groups involved in this initiative (users, managers of some partner non-profit organisations and managers at Ecoembes).

1 Introduction

Socially responsible consumption is where consumers take into account traditional, environmental, social and ethical criteria from the moment they decide to buy a product to its disposal, and as such, any available recycling options contribute to this type of consumption.

M. M. Palacios-González (✉) · M. S. Janita-Muñoz
Universidad de Extremadura, Badajoz, Spain
e-mail: mapalaciosg@unex.es; msjanita@unex.es

E. C. Bianchi et al. (eds.), *Responsible Consumption and Sustainability*, Springer Business Cases, https://doi.org/10.1007/978-3-031-30742-3_5

Indeed, one of the Sustainable Development Goals (SDGs), number 12, devised by the United Nations (UN), is "Responsible Production and Consumption". This goal is based on "doing more and better with less", in other words, increasing the efficiency of available resources and encouraging healthy lifestyles. This goal is based on specific targets, one of which specifically mentions the process of waste recycling and reuse: "By 2030, substantially reduce waste generation through prevention, reduction, recycling and reuse" (UN, 2022).

In this context, the European Union presented the Circular Economy Action Plan (CEAP) in 2020 as part of its sustainable growth strategy. The measures contained in the Plan include the reduction of excessive packaging, new mandatory requirements on recycled content (with a special focus on microplastics, bioplastics and biodegradable plastics), preventing waste generation and, where waste already exists, transforming it into high-quality secondary raw material market (European Commission, 2020).

According to the latest data available, the average waste recycling rate (glass, plastic, paper/card, metal, wood and others) in the European Union rose to 64.4%; meanwhile, in Spain, this figure rose to 69.6% (Eurostat, 2022).

Furthermore, the website of Ecoembes,[1] the non-profit organisation responsible for coordinating light household packaging waste recycling in Spain, tracks the evolution of domestic waste recycling (plastics, paper/card, metal, cartons and wood) over recent years in Spain. In 2019, this figure was 1,505,661 tonnes of waste, being slightly lower during the year of the pandemic (in 2020, it was 1,490,283 tonnes of waste), whereas in 2021, the figure exceeded the pre-pandemic levels and reached 1,570,513 tonnes of waste (Ecoembes, 2022).

The tireless efforts by Ecoembes to contribute to reducing the environmental impact of waste and to its circularity have led it to set up the first Spanish Circular Economy laboratory, also a pioneer in Europe in the area of packaging waste and its recycling, *TheCircularLab.*

One of the most high-profile initiatives from the innovation centre is RECICLOS. It consists of a Deposit Return Scheme (DRS) that uses mobile technology to encourage responsible environmental consumer behaviour by offering "rewards". Every time a consumer deposits waste packaging in the designated collection container, they register the deposit on the RECICLOS systems. They can then claim sustainable, eco-friendly prizes beneficial to all.

2 Ecoembes: Union Means Strength

Ecoembes is a non-profit organisation founded in Spain in 1996 by packaging businesses, distributors and manufacturers, as a cooperative and effective response to the requirements of Law 11/97, of 24 April, on packaging and packaging waste.

[1]The information appearing in this paragraph was taken from the Ecoembes website (https://www.ecoembes.com/es).

This regulation was transposed from the European Parliament and Council Directive 94/62/EC of 20 December 1994 on packaging and packaging waste. Both legal provisions require packaging businesses, distributors and manufacturers to take responsibility for the environmental impact of their products and contribute to the recovery of the raw material, which represents an important step towards a circular economy in Spain (Ecoembes, 2022).

Its mission is to develop strategies, processes, actions and projects linked to the different stages of the life cycle of packaging, in conjunction with those responsible for each stage, with the ultimate aim being to contribute to reducing the environmental impact and to the circularity of this type of waste (Ecoembes, 2022).

With the aim of developing this function, its management model is based on public-private partnership involving businesses, public entities and citizens. This model is based on efficiency, transparency and technological innovation principles, with the aim of achieving extremely ambitions recycling objectives (Ecoembes, 2022).

The Ecoembes value proposition rests on the following pillars (Ecoembes, 2020):

- Promote recycling and circularity of packaging and ensure that the legal objectives for recycling are met, with the aim of attaining the maximum levels of selective collection of household packaging.
- Contribute to reducing the use of raw materials and the consumption of water and energy.
- Help and incentivise citizens to recycle more and better, becoming a point of reference for recycling education, awareness and data transparency.
- Strengthen society engagement, deepening public-private partnerships, building trust and sharing values with citizens, companies, public authorities and civil society.
- Guarantee sustainability and efficiency in its activity, working jointly with administrations to ensure the quality of the selective collection service.
- Promote innovation and offer businesses the knowledge, tools and resources they need in order to produce more sustainable packaging using eco-design.

In addition to directly contributing to SDG number 12, the organisation also pursues nine other SDGs, including quality education, gender equality and reduced inequalities, with the aim of shaping a more circular, prosperous, green and inclusive future (Ecoembes, 2022).

Ecoembes is a member of several European organisations involved in the circular economy and recycling: the Packaging Recovery Organisation Europe (PRO Europe), whose main role is as the licensor of the "Green Dot" trademark predominantly in Europe; the Extended Producer Responsibility Alliance (EXPRA), which focuses on national Extended Producer Responsibility (EPR) in packaging; and the European Circular Economy Stakeholder Platform (ECESP), which promotes information sharing on the circular economy, strategies and good practice between the different countries of the European Union.

PRO Europe is the oldest of these organisations, founded in 1995, and responsible for coordinating European plans for packaging and packaging waste recovery and packaging waste that carries the "Green Dot" trademark as a financing symbol. PRO Europe members are responsible for organising the efficient implementation of national packaging collection and recovery systems. "The Green Dot" has become a way of developing manufacturer and distributor responsibility in many countries, given that when packaging, distribution or manufacturing companies use the trademark on their packaging; it shows that they have made a financial contribution to a packaging recovery company in their country, who is a member of PRO Europe, for them to carry out their work. Current PRO Europe members are as follows: ARA (Austria), Fost Plus (Belgium), Exopak D.O.O. (Bosnia-Herzegovina), ECOPACK (Bulgaria), EKO OZRA (Croatia), Green Dot Cyprus (Cyprus), EKO-KOM (Czech Republic), ETO (Estonia), CITEO (France), Der Grune Pünkt (Germany), Hellenic Recovery Recycling Corporation (Greece), ÓKO-PANNON (Hungary), REPAK (Ireland), TAMIR (Israel), JSC LATVIJAS ZALAIS PUNKTS (Latvia), ZALIASIS TASKAS (Lithuania), VALORLUX (Luxembourg), GREENPACK (Malta), PAKOMAK (North Macedonia), Afvalfonds Verpakkigen (Netherlands), GRONT PUNKT (Norway), REKOPOL (Poland), Sociedade Ponto Verde (Portugal), ECO-ROM AMBALAJE (Romania), SEKOPAK (Serbia), ENVI-PAK (Slovakia), SLOPAK (Slovenia), Ecoembes (Spain), FTI AB (Sweden), CEVKO (Turkey) and VALPAK (United Kingdom) (PRO Europe, 2022).

Another key European organisation is EXPRA, whose president in 2022 was the CEO of Ecoembes, Óscar Martín Riva. EXPRA is a European non-profit organisation founded in March 2013, whose mission is to uphold Extended Producer Responsibility relating to packaging. In other words, they promote environmental and prevention responsibilities by those responsible for launching packaged products onto the market, right from the design phase to the point at which they become waste and their subsequent management. The organisation represents the interests of all its member organisations, acting as a common political platform. The members of EXPRA operate under the principles of transparency and maximum environmental, economic and social efficiency. EXPRA is made up of 26 members: Fost Plus (Belgium), Valipac (Belgium), EKOPAK (Bosnia and Herzegovina), ECOPACK (Bulgaria), Éco Entreprises Québec (Canada), Green Dot Cyprus (Cyprus), EKO-KOM (Czech Republic), ETO (Estonia), RINKI (Finland), HERRCO (Greece), Öko Pannon (Hungary), Icelandic Recycling Fund (Iceland), TAMIR (Israel), CONAI (Italy), VALORLUX (Luxembourg), PAKOMAK (North Macedonia), Greenpak (Malta), Afvalfonds Verpakkingen (Netherlands), Green Dot Norway (Norway), Eco-Rom Ambalaje (Romania), Envipak (Slovakia), SLOPAK (Slovenia), Ecoembes (Spain), Ecovidrio (Spain), FTI (Sweden) and CEVKO (Turkey) (EXPRA, 2022).

Furthermore, an initiative from the European Commission brought about the creation of the ECESP, founded in March 2017 by the European Economic and Social Committee (EESC) for the purpose of sharing circular economy strategies, knowledge and best practice between member states. This knowledge exchange takes place via a virtual platform and a coordination group made up of 24 members.

The coordination group is made up of both public and private sector businesses and also non-governmental organisations (ECESP, 2022). Ecoembes was chosen to join this coordination group. Its involvement in this platform is focused on raising the profile of innovation within the circular economy developed in *TheCircularLab*, as well as other aspects of recycling. *TheCircularLab* is the first circular economy laboratory in Spain and is also a pioneer in Europe in the area of packaging waste and its recycling (Ecoembes, 2022).

3 Circular Economy Innovation: *TheCircularLab*

TheCircularLab is the first European innovation centre focused on the circular economy and was founded in Logroño (La Rioja, Spain) in 2017.

The centre aims to study, conceive, test and apply best practices relating to packaging in a real environment. These procedures apply to all phases of packaging life cycle from conception to its reintroduction into the consumption cycle through new products. To this end, collaboration with citizens, businesses and public administrations is key (Thecircularlab, 2022).

Table 1 shows each of the areas of innovation in which the centre works, as well as the current innovation projects being undertaken in each of them.

As listed in Table 1, RECICLOS is one of the innovative projects to come out of *TheCircularLab*.

4 The RECICLOS Initiative

Although recycling rates for household packaging in Spain are encouraging, Ecoembes aims to keep coming up with innovative ideas to adapt to new citizen behaviours which promote recycling habits and also socially responsible consumption.

RECICLOS was developed through an initiative of TheCircularLab, Ecoembes' own innovation centre. It consists of a Deposit Return Scheme (DRS) in Spain that rewards responsible environmental behaviour by inhabitants, through the recycling of cans and plastic drinks bottles (Thecircularlab, 2022). The key objective of the initiative is to increase the rates of selective waste packaging collection (Ecoembes, 2020).

The initiative involves the modernisation of existing infrastructures (yellow containers) which inhabitants are familiar with, which to a large extent encourages their use, and also the installation of new specially designed machines. This modernisation has made possible innovation using mobile technology and a reward system with sustainable incentive points. With this scheme, the RECICLOS initiative promotes a habit of recycling through the use of mobile technology and a reward system (Thecircularlab, 2022).

This initiative links directly with the United Nations' definition of Smart Cities. According to the UN, a smart city *uses ICTs and other means to improve quality of*

Table 1 TheCircularLab innovation areas and projects

Innovation areas	Innovation project	
	Name	Brief description
Eco-design	Pack CD	An innovative tool that helps companies predict and modify the environmental impact of their containers before they are manufactured
	Observatory of the Packaging of the Future	A platform of knowledge about packaging, which collects and analyses news and trends that appear worldwide
	Bio-Bio Plastic	A bio-based material obtained from 100% decomposing organic material, which is recyclable, compostable and biodegradable
Citizen science	RECICLOS	A new return and reward system that, through mobile technology, offers incentives to reward the environmental commitment of inhabitants who recycle cans and plastic bottles in yellow bins or special machines
	Circular Talent Labs	A training and employment programme that supports and encourages young multidisciplinary talent through projects related to the circular economy
Smart waste	Smartwaste	An innovative technological and information and management platform that applies artificial intelligence and big data in order to improve the efficiency of the collection, selection and recycling of packaging processes
Entrepreneurship	GoCircular	A new open innovative community for start-ups and entrepreneurs in the circular economy. It aims to generate knowledge sharing, promote innovative solutions and provide guidance for its implementation
	Circular design	A methodology which enables multidisciplinary profiles to be given training on the circular economy while at the same time providing answers to challenges to improve waste packaging

Source: Compiled based on the TheCircularLab (2022)

life, efficiency of urban operation and services, and competitiveness, while ensuring that it meets the needs of present and future generations with respect to economic, social, environmental as well as cultural aspects (UN, 2022).

4.1 How RECICLOS Works

Below are detailed instructions for consumers on how to use RECICLOS (RECICLOS, 2022; Thecircularlab, 2022):

Fig. 1 RECICLOS app home page. Source: RECICLOS (2022)

1. Download the RECICLOS app, available from the Google Play Store and Appstore, and sign up (Fig. 1).
   The technological evolution of RECICLOS incorporates innovation in the form of blockchain,[2] artificial intelligence and image recognition that recognises images and QR codes.[3] By using this technology, the app is able to strengthen the links between inhabitants and collection containers (both the yellow

[2] Blockchain is a shared, immutable ledger for recording transactions, tracking assets and building trust. Assets could be tangible (houses, cars, cash, land) or intangible (intellectual property, patents, copyrights, brands) (IBM, 2022).

[3] A quick response code or QR code is a two-dimensional version of the barcode able to convey a wide variety of information almost instantly with the scan of a mobile device (QR code generator, 2022).

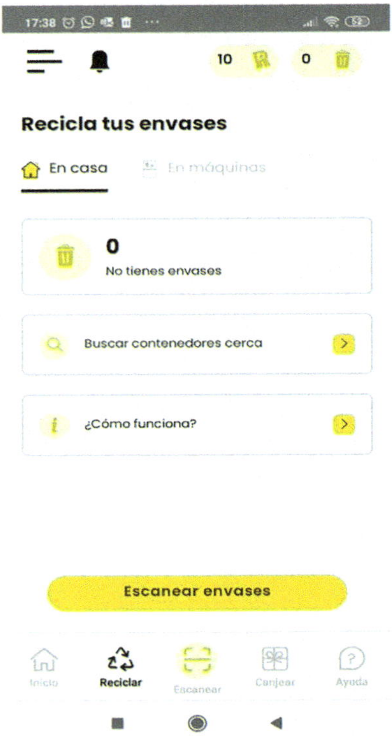

Fig. 2 RECICLOS APP scanning page. Source: RECICLOS (2022)

containers and the specially designed machines) by means of mobile phones, to make recycling a habit that is rewarded.

2. Scan the drink cans and plastic bottles you have consumed (Fig. 2).
3. Deposit the cans and bottles in the yellow container or in the RECICLOS machines. If you use the yellow containers, the container must have a QR code. Once you have deposited the cans and bottles, you must scan the QR code on the container (Fig. 3).
   These containers were first implemented in several municipalities in Catalonia as a pilot project. Following a good response from inhabitants, they were installed in several other municipalities of this Autonomous Community. They are also gradually being installed in all the Spanish Autonomous Communities, so that at least one municipality in each Community will have these types of containers.

Cans and bottles can also be deposited in specially designed machines, similar to vending machines, but in reverse (Fig. 4). Once the cans and bottles have been deposited, the QR code on the machine must be scanned. The main objective of the machines is to encourage recycling outside the home.

In Spain, the region where the installation of specially designed machines was installed was Catalonia in June 2019. However, there are now machines in several Communities, such as Andalusia, Aragon, Castile and León, Catalonia,

Fig. 3 Yellow container deposit. Source: RECICLOS (2022)

Fig. 4 Machine deposit. Source: RECICLOS (2022)

Community of Madrid, Valencian Community, Galicia, Balearic Islands and La Rioja. Indeed, these machines have been installed in different public places such as shopping centres (Zenia in Orihuela, Los Arcos and Alcampo Tamarguillo in Seville and Ponte Vella in Ourense, Camino de la Plata in Burgos, Parque Rioja in

Logroño and La Dehesa in Alcalá de Henares), train stations (Ferrocarrils de Cataluña), entertainment facilities (Science Museum, Umbracle and L'Oceanogràfic in Valencia), hospitals (Hospital Sanitas La Zarzuela), seaports (Balearia marine terminals in Valencia and Denia), universities (Complutense University of Madrid), etc.

Ecoembes plans to keep growing the RECICLOS initiative, both through yellow containers with QR codes and specially designed machines, gradually reaching all municipalities and public areas in Spain.

4. Get 1 RECICLOS point for each container deposited in the containers or machines. It is worth noting that the system has a weekly limit for cans and plastic bottles that inhabitants can get recycling points for, the reason being that the Recycling Reward System aims to encourage responsible and rational consumption of this type of packaged product.
5. Redeem the RECICLOS points to enter prize draws or make donations to social or environmental projects and sustainable mobility initiatives (Fig. 5).

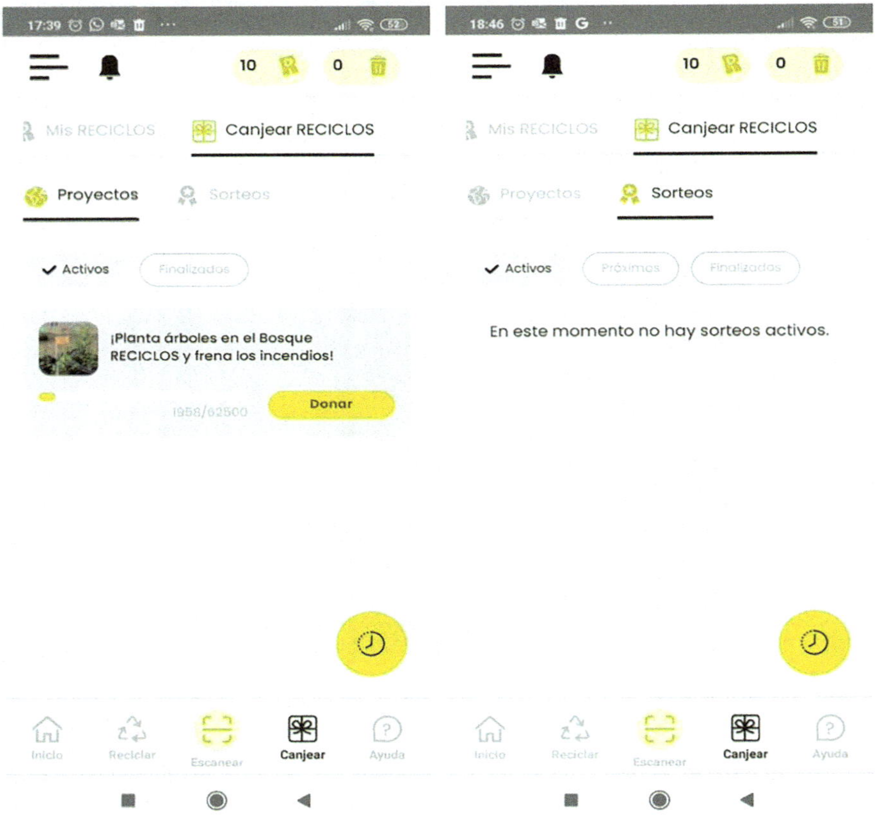

Fig. 5 RECICLOS app rewards and donations page. Source: RECICLOS (2022)

4.2 RECICLOS Results in Spain

The results achieved by the application in Spain are collated on the Ecoembes website (Ecoembes, 2022). In July 2022, the RECICLOS technology was being used with over 24,000 yellow containers distributed in municipalities across Spain. There have also been 180 specially designed machines installed in a range of public spaces such as sports centres, shopping centres, hospitals and universities.

The number of RECICLOS users in Spain now stands at over 200,000, taking into account both collection infrastructures.

These users have entered numerous prize draws, and a range of different prizes have been won, including electric bikes, electric scooters, tablets and products from the "kilometre 0" shopping basket which promote sustainable mobility, fair trade, etc. In real terms, by July 2022, over 21,000 prizes had been awarded. Ecoembes funds these prizes.

The RECICLOS initiative also encourages users to donate the points they receive for recycling drinks cans and plastic bottles to different social or environmental projects as well as sustainable mobility initiatives. To this end, Ecoembes has signed agreements known as collaboration agreements with different public administrations and non-governmental organisations. These agreements establish, among other things, which collaboration project it involves as well as the volume of donations to be reached. For each project, the points earned through RECICLOS correspond to a certain amount of euros, which is not the same for every project. In July 2022, the amount of points that had been donated had reached over 7.8 billion points.

These points have supported over 320 social or environmental projects and sustainable mobility initiatives.

Social projects included donations to the Red Cross, to food banks in different cities and to local non-profit organisations which help vulnerable residents. For example, masks and alcohol gels were donated by the residents of Bullas (Murcia), a donation to the Asociación Canaria del Trastorno del Espectro del Autismo (APANATE) by residents of Granadilla de Abona (Tenerife), a donation from Spanish citizens to the Red Cross for those in crisis, including those affected by the Cumbre Vieja volcano (La Palma) and the war in Ukraine.

Donations have also been made to environmental projects and sustainable mobility initiatives, including cleaning up Monte Ulía (San Sebastián), the recovery of green urban spaces in Santa Marta (Santiago de Compostela), improvements to Parque Pinar Ranilla (Seville) and the installation of bicycle parking spaces in Granollers and Figueres (Catalonia). The planting of the RECICLOS forest is another initiative, aimed at tackling desertification and restoring forested areas.

In the future, the number of RECICLOS users in Spain is predicted to rise, given plans to gradually increase the number of containers with QR codes and the number of specially designed machines in different municipalities and public spaces in Spain. There are also plans to install "intelligent yellow containers" fitted with technology which identifies the waste deposited in them. There are an ever-increasing number of inhabitants getting involved in recycling drinks cans and plastic bottles and claiming prizes or donating their points to a large number of social or environmental projects and sustainable mobility initiatives.

4.3 Testimonies from the RECICLOS Initiative in Spain

Finally, in Table 2, there are testimonies from people who have participated in the initiative, from the organisers Ecoembes and also from some of the NGOs responsible for social and environmental causes.

Table 2 RECICLOS testimonies

User testimonies
"It is not just about the prizes, it's also because recycling is FUNDAMENTAL" (Camila)
"An electric scooter in exchange for recycling some plastic bottles. . .BAM!" (Jeffrey)
"Recycling is fundamental and now you can win incredible prizes" (Mercè)
"Recycle with the RECICLOS app and win sustainable prizes" (Laura)
"A brilliant initiative from Ecoembes. Now there are no excuses not to recycle" (Esteban)
In response to the food donations received as a result of the RECICLOS initiative, the president stated that "It has been extremely rewarding for the Fundación Banco de Alimentos de Sevilla to be involved, with Ecoembes, in such an exciting and necessary project such as RECICLOS, as raising awareness about caring for our environment and encouraging recycling is everyone's responsibility. We are also very grateful for the support for our organisation, which has enabled us to stock our warehouses with over 7500 kilos of food, helping to meet the needs of families in need in the province of Seville" (Agustín Vidal-Aragón de Olives, the president of the Seville Food Bank)
Regarding the different community actions aimed at improving the environment that have been made possible through this initiative, the president commented that "The involvement of the Red Cross in projects initiated by RECICLOS has allowed us to exponentially increase our awareness of responsible public behaviour relating to our environment. The partnership with RECICLOS has enabled us to increase the scope of Red Cross projects, such as the Promotion of Environmentally Responsible Habits and Conduct and the Prevention of Forest Fires in "La Devesa" of the Albufera National Park. This collaboration has placed value on the role of volunteers dedicated to the conservation and protection of the environment" (Nieves Dios, local president of the Red Cross in Valencia)
The director places value on the unity and solidarity demonstrated by citizens highlighting that "We are extremely proud of how Spanish citizens have got behind social and environmental causes by using their RECICLOS points. It is no longer only about engagement with the Deposit Return Scheme, it is so much more than that: recycling is becoming a way to care not only for the environment, but also for those in needs, as some 100,000 users have demonstrated" (Nieves Rey, Communications and Marketing Director of Ecoembes)

Source: Complied by Reciclos (2022) and Ecoembes (2022)

5 Conclusions

Nowadays, people have information on the social and environmental consequences of their consumption habits, meaning that when buying a product or service, not only do they have to take into account traditional criteria (price, quality, etc.) but they also have to think about the repercussions that their purchase has on society and the environment.

Among those repercussions, recycling of waste packaging was found to lower its impact on the environment. According to data from Eurostat and Ecoembes (the non-profit organisation responsible for coordinating light household packaging waste recycling in Spain), the recycling rates in Spain are high.

With the aim of maintaining this trend, RECICLOS was born, offering rewards for recycling through prizes or donations to social or environmental projects and sustainable mobility initiatives.

In July 2022, the number of users in Spain exceeded 200,000. Users deposit drinks cans and plastic bottles in one of the 24,000 yellow containers with QR codes or in one of the 180 or more specially designed machines located in public spaces. Thanks to this initiative, users have won different prizes and have also contributed to over 320 social or environmental projects and sustainable mobility initiatives.

Users are pleased to be contributing to caring for the environment as they believe it to be something “fundamental”. For their part, the managers of some of the partner organisations have appreciated Ecoembes’ commitment to social and environmental causes.

Ecoembes themselves have wanted to thank all the users for taking part in this initiative and for helping them care not only for the environment but also for those most in need. In view of this, the organisation predicts an increase in the number of users of the initiative in the coming years, given that there are plans to gradually roll out the scheme to all municipalities and public spaces in Spain.

Lastly, it must be noted that RECICLOS has a positive impact on the environment through its promotion of depositing waste packaging in the correct containers for recycling. What is more, these deposits give users a chance to obtain points with which they can enter draws to win sustainable prizes or make donations to social or environmental projects and sustainable mobility initiatives. Therefore, there is no doubt that the RECICLOS initiative is an action that encourages responsible social consumption and directly contributes towards the 12th SDG.

6 Discussion Questions

Question 1: Can you name any initiatives similar to RECICLOS in your own country? Below, indicate if they are linked to a non-profit organisation or not, what their objective is and how the initiative works.

Question 2: Analyse the results obtained by the initiative named in question 1. What is the perception of both the users of the initiative and the non-profit organisations which benefit from the social cause supported by it?

Question 3: Is it appropriate or not that these initiatives have emerged as a result of legal regulations imposed by public administrations (as is the case with RECICLOS, an initiative promoted by the non-profit organisation in response to European Parliament and Council Directive 94/62/EC of 20 December 1994 on packaging and packaging waste and Law 11/97, of 24 April, on Packaging and Packaging Waste in Spain)? Discuss this question and then share ideas with the rest of the class.

Acknowledgements The authors acknowledge the financial support provided by the Junta de Extremadura and FEDER under the grant GR21078.

The authors of this paper would like to thank all the staff members at Ecoembes who have contributed, in particular Laura Ramos, Silvia López, Soledad Albarrán and Iñaki Campo.

Appendix: Teaching Notes for the Case

Synopsis

Broadly speaking, socially responsible consumption is where consumers take into account traditional, environmental, social and ethical criteria from the moment they decide to buy a product to its disposal, by engaging in any available recycling options.

According to a report on the recycling of packaging waste published by Eurostat (2022) concerning the year 2019, the average recycling rate in the European Union rose to 64.4%; meanwhile, in Spain, this figure rose to 69.6%. In order to increase this rate, Ecoembes, the non-profit organisation responsible for coordinating light household packaging waste recycling in Spain, has launched the RECICLOS initiative.

The initiative consists of a Deposit Return Scheme (DRS) that uses mobile technology to encourage responsible environmental consumer behaviour by offering "rewards". Every time consumers deposit waste packaging in the designated collection container, they register the deposit on the RECICLOS systems. They can then either claim sustainable, eco-friendly prizes beneficial to all or donate the reward to social or environmental projects and initiatives linked to sustainable mobility. The RECICLOS initiative can therefore be considered as an action that encourages socially responsible consumption.

Here, the RECICLOS initiative will be discussed, and the key results achieved since its launch will be explained.

Potential Audience and Instructor's Material

This case has been developed for use in the framework of socially responsible consumption and packaging recycling seminars in particular.

The potential audience for this chapter is twofold:

- Undergraduate students pursuing degrees in business administration or marketing
- Graduate students pursuing MBA and masters in marketing or executive programs and seminars (about socially responsible consumption in general and packaging recycling in particular).

Learning Objectives

The general objective of this paper is to raise the profile of the RECICLOS initiative devised by the non-profit organisation responsible for coordinating light household packaging waste recycling in Spain, Ecoembes, through its laboratory *TheCircularLab.*

It also seeks to achieve the following more specific objectives:

- Reflect on the change that is currently taking place in consumption habits, giving rise to "socially responsible consumption" and its links with waste packaging recycling, as well as understanding the importance of this type of consumption in the future strategies of international and European organisations.
- Analyse waste recycling data in the European Union and Spain.
- Learn more about the non-profit organisation Ecoembes, responsible for coordinating light household packaging waste recycling in Spain, its mission, pillars and links with other similar organisations across Europe.
- Study the RECICLOS initiative, its operation, main achievements and the perceptions of the various interest groups involved in this initiative (users, managers of some partner non-profit organisations and managers at Ecoembes).

Time Frame for Class Discussion

This case can be taught in one session of 60–100 minutes, depending on the number of students participating in the discussion activities, as well as the previous work that students have done before class.

Suggested Discussion Questions

1. Can you name any initiatives similar to RECICLOS in your own country? Below, indicate if they are linked to a non-profit organisation or not, what their objective is and how the initiative works.
2. Analyse the results obtained by the initiative named in question 1. What is the perception of both the users of the initiative and the non-profit organisations which benefit from the social cause supported by it?
3. Is it appropriate or not that these initiatives have emerged as a result of legal regulations imposed by public administrations (as is the case with RECICLOS, an initiative promoted by the non-profit organisation in response to European

Parliament and Council Directive 94/62/EC of 20 December 1994 on packaging and packaging waste and Law 11/97, of 24 April, on Packaging and Packaging Waste in Spain)? Discuss this question and then share ideas with the rest of the class.

Analysis

Answer to question 1: *Can you name any initiatives similar to RECICLOS in your own country? Below, indicate if they are linked to a non-profit organisation or not, what their objective is and how the initiative works.*

If this activity is carried out in the classroom, the number of students should be divided into groups of four.

Each group will identify an initiative similar to RECICLOS in their own country. Students will be asked to answer the following questions:

- Initiative name.
- Name of the non-profit organisation or any other type of organisation the initiative is linked to. If a non-profit organisation is not involved, the legal status of organisation should be identified.
- Initiative objective: encourage waste recycling, support underprivileged groups, etc.
- How the initiative works: the full details of the steps that must be taken by someone participating in the initiative must be explained.

When the task has been completed, the members of each group will share ideas and create a presentation. Each group will present the initiative studied to the rest of the students.

Answer to question 2: *Analyse the results obtained by the initiative named in question 1. What is the perception of both the users of the initiative and the non-profit organisations which benefit from the social cause supported by it?*

If this activity is carried out in the classroom, the number of students should be divided into groups of four.

Each group should provide as much detail as possible about the results the initiative named in question 1 has achieved in their country and the perception of both the users of the initiative and the non-profit organisations which benefit from the social cause supported by it.

When the task has been completed, the members of each group will share ideas and create a presentation. Each group shall present the results of the initiative to the rest of the class.

Answer to question 3: *Is it appropriate or not that these initiatives have emerged as a result of legal regulations imposed by public administrations (as is the case with RECICLOS, an initiative promoted by the non-profit organisation in response to European Parliament and Council Directive 94/62/EC of 20 December 1994 on packaging and packaging waste and Law 11/97, of 24 April, on Packaging and*

Packaging Waste in Spain)? Discuss this question and then share ideas with the rest of the class.

If this activity is carried out in the classroom, the number of students should be divided into groups of four.

Each group should discuss the appropriateness or not that these initiatives have emerged as a result of legal regulations imposed by public administrations (as is the case with RECICLOS, an initiative promoted by the non-profit organisation in response to European Parliament and Council Directive 94/62/EC of 20 December 1994 on packaging and packaging waste and Law 11/97, of 24 April, on Packaging and Packaging Waste in Spain).

When the task has been completed, the members of each group will share ideas and create a presentation. Each group will present their arguments to the rest of the class, and then ideas will be shared with the whole class.

Open responses are encouraged for this task, depending on the opinions of each group. Three scenarios are possible. Firstly, it may be thought that it is appropriate that these types of initiatives have emerged as a result of legal regulations imposed by public administrations. Secondly, it may be thought that the origin of these initiatives is not relevant; what is important is that they exist and foster support for different social or environmental causes. Finally, it may be thought that it is not appropriate that these initiatives have emerged as a result of legal regulations; instead, they should have already been encouraged prior to the regulations being imposed. In all the different scenarios, the students should explain the reasons or grounds for each of the arguments.

References

Ecoembes. (2020). *Ecoembes: more digital, more circular*. Executive summary 2020.

Ecoembes. (2022). *Official information from the website*. Consulted July 21, 2022.

ECESP. (2022). *Official information from the website*. Consulted July 21, 2022.

European Commission. (2020, March 11). *Changing how we produce and consume: New Circular Economy Action Plan shows the way to a climate-neutral, competitive economy of empowered consumers*. Press Release.

Eurostat. (2022). *Recycling rates for packaging waste*. Consulted July 13, 2022.

EXPRA. (2022). *Official information from the website*. Consulted July 21, 2022.

IBM. (2022). *Official information from the website*. Consulted July 21, 2022.

PRO Europe. (2022). *Official information from the website*. Consulted July 21, 2022.

QR code generator. (2022). *Official information from the website*. Consulted July 21, 2022.

RECICLOS. (2022). *Official information from the website*. Consulted July 21, 2022.

Thecircularlab. (2022). *Official information from the website*. Consulted July 21, 2022.

United Nations (UN). (2022). *Official information from the website*. Consulted June 27, 2022.

LessPack: A Project Helping Planet Sustainability

M. Manuela Palacios-González and Antonio Chamorro-Mera

Learning Objectives

- Reflect on the impact generated by packaging waste on planet sustainability.
- Understand the importance of innovation and eco-design in new packaging to prevent waste generation and to facilitate the recycling and reuse of packaging already produced.
- Learn about the key motives and barriers to participation in the initiative.

1 Introduction

Regrettably, the frequency of natural and human disasters seen in our planet is ever increasing. These occurrences compromise the planet's capacity to retain its natural resources long term and, therefore, life as we know it. As a result, society in general, as well as businesses and especially consumers, is increasingly aware of the need for changes to current production methods and consumption in order to make the planet more sustainable.

In the same vein as Sustainable Development Goal 12, the circular economy action plan adopted in 2020 by the European Union focuses on design and production fitting into a circular economy. The measures covered in the plan include ensuring less waste is produced. This firstly involves reducing waste and secondly producing high-quality secondary raw materials that can be integrated into an effective secondary raw materials market (European Commission, 2020).

Statistical data on waste generation show that in 2020, the amount of municipal waste generated per person in the EU amounted to 505 kg. As regards differences in the volume of waste generated between 1995 and 2019, the EU saw an increase, that is to say, waste generated in this period increased by 8.2% (Eurostat, 2022). A significant proportion of municipal waste is made up of packaging waste. Of that

M. M. Palacios-González (✉) · A. Chamorro-Mera
University of Extremadura, Badajoz, Spain
e-mail: mapalaciosg@unex.es; chamorro@unex.es

E. C. Bianchi et al. (eds.), *Responsible Consumption and Sustainability*, Springer Business Cases, https://doi.org/10.1007/978-3-031-30742-3_6

Fig. 1 Plastic waste in nature. *Source: Royalty-free image from* pixabay.com

waste, plastic containers are the most challenging due to the length of time it takes for them to decompose when they are dumped in nature (Fig. 1). In real terms, each person in the EU generated an average of 34.4 kg of plastic waste, which is 24% more than generated a decade before. Furthermore, only 41% of these products are recycled (Eurostat, 2022).

With the aim of contributing to improvements in the reduction of packaging waste and its recycling or recovery, the non-profit organisation responsible for coordinating light household packaging waste recycling in Spain, *Ecoembes*, the Asociación Vertidos Cero and company *Recircular* have developed a collaborative project known as *LessPack*.

The project was started in 2019 and has two main objectives: promote initiatives that reduce the generation of packaging waste, and identify opportunities aligned with both the many legislative requirements and consumer demand for more sustainable consumption options.

This document describes the organisations involved in this initiative, as well as looking at what it involves, the actions carried out and the recognition it has received to date.

2 Reducing Waste Through the Strength of Collaboration

This section describes each of the organisations involved in this collaborative project initiative.

2.1 Ecoembes[1]

Ecoembes is the non-profit organisation responsible for coordinating light household packaging waste recycling in Spain and helps companies to integrate eco-design into the production of their packaging to encourage a reduction in the generation of packaging waste.

In order to carry out its work, Ecoembes partners with public administrations, companies and the public. By working with Ecoembes, companies can satisfy current European Union obligations to take responsibility for waste recovery relating to their packaged products aimed at the final consumer. Companies pay a fee to Ecoembes for each packaged product they put on the market. Subsequently, Ecoembes partners with local governments to establish a system of selective collection street containers which inhabitants can use to deposit different types of packaging waste. Once collected, waste is sold to specialist recycling companies. The revenue from the sales and fees funds the system.

In addition to this main activity, Ecoembes also develops different projects in other areas of activity, with the aim of reducing waste generation or waste recovery, as shown in Table 1.

2.2 Asociación Vertidos Cero[2]

The *Asociación Vertidos Cero* (translated as Zero Waste Disposal Association) is a non-profit organisation created by a group of professionals in response to social concerns and the impact of waste on the environment, and to improve the quality of life for the whole of society. The ultimate aim of this association is to achieve solidarity between current and future generations by promoting an economic and production model that is environmentally friendly and in line with the circular economy.

The *Asociación Vertidos Cero* operates as a platform for research, innovation, training and raising awareness about the issue of waste.

Table 2 shows the non-profit organisation's current projects and a brief description of them.

[1] The information in this sub-section was taken from the official website of Ecoembes: https://www.ecoembes.com/es (Ecoembes, 2022).

[2] The information in this sub-section was taken from the official website of the *Asociación Vertidos Cero*.

Table 1 Ecoembes projects up to 2022

Area	Project name	Brief project description
Environment	Libera (liberate)	Aims to raise awareness about the problem of littering,[a] prevent recurrence and involve all sectors of society in finding solutions. This project was devised alongside SEO/BirdLife.
	Upcycling the oceans	Aims to remove marine debris from the bottom of the oceans and transform it into top quality yarn. In this project, Ecoembes partners with the ECOALF (Spanish sustainable fashion brand).
	Camino del reciclaje (the recycling way)	Aims to keep the Camino de Santiago litter free. Ecoembes partners with EcoPeregrinos and EcoAlbergues in this project.
Society	La música del reciclaje (the music of recycling)	Aims to help children at risk of social exclusion through music education. It also promotes fundamental values, reinforces the importance of recycling and also teaches participants how to make their own recycled instruments
	Terceros en edad, primeros en reciclar (late in life, first to recycle)	Specially aimed at the elderly, encouraging them to recycle more and better through different leisure activities, nature trips and many other group activities
	Reciclar Para cambiar vidas (recycling to change lives)	A social inclusion project through employment to improve the employability of vulnerable people with a view to their employment in the recycling industry
Education	Naturaliza	Aims to boost environmental education in all schools and encourage more interaction with nature through promoting hands-on environmental learning
	Amarillo, Verde y Azul (yellow, green and blue)	Aims to encourage responsible consumption, reuse and recycling through an extensive and fun blog, a place where teachers and their students are encouraged to protect the environment
	Recursos educativos (educational resources)	Provides the educational community with a broad range of materials, games, posters and other teaching tools, as well as suggestions for outdoor activities, to encourage environmental awareness in the classroom
	Actividades de Educación especial (special education activities)	Programmes and materials for children with intellectual impairments and autism spectrum disorder (ASD)
Innovation	TheCircularLab	The creation of an open innovation centre specialised in the circular economy, whose activity is focused on the study, trial and

(continued)

Table 1 (continued)

Area	Project name	Brief project description
		development of best practices in the packaging sector and subsequent recycling
	TheCircularCampus	The creation of a space for knowledge and further education on circular economy and packaging to contribute to environmental sustainability by achieving packaging circularity
Other areas	Ecohostelero	Aimed at the HORECA channel and all hoteliers who wish to make their businesses environmentally friendly. Training, proper recycling of all waste produced and material management are its major goals
	LessPack	Aims to encourage less packaging waste production
	Zero cabin waste	A pioneering project in the airline business; it aims to recover 80% of the waste produced in aircraft cabins

Source: Compiled by Ecoembes (2022)
[a]Littering here refers to the problem of waste generated by humans being discarded in nature

2.3 Recircular[3]

Recircular is a company specialising in the buying and selling of resources via an online platform. The services it offers can be grouped into three main categories: waste recovery platform, impact measuring and innovation projects.

Its principal service involves the use of a digital platform where companies can sell resources, they no longer use waste, by-products and excess materials and stock material, so that other companies can buy them and give them a second life. This service acts as a point of connection between potential sellers and buyers. The *Recircular* platform also provides information to its clients on the possible reuse and recycling options for different resources. Lastly, if a certain resource is not available on the platform, companies can submit a request via the platform and will then be informed if there are any companies that could potentially supply that resource.

Impact measuring refers to a methodology developed by *Recircular* in collaboration from the Basque Circular Hub technology centre. It is used to calculate company's environmental and social benefits generated through the recovery of resources. Currently, we can measure the following indicators: CO2 emissions, water and energy consumption and people's life expectancy. Nevertheless, *Recircular* plans to increase the number of socioeconomic indicators in its study,

[3]The information in this sub-section was taken from the official website of *Recircular*: https://recircular.net (Recircular, 2022).

Table 2 *Asociación Vertidos Cero* projects up to 2022

Project name	Brief project description
Corazón de Corcho (Cork heart)	Aims to reuse waste wine corks as a new resource to be reintegrated into the production chain, an example of the circular economy. This initiative was devised in conjunction with fashion designer Agatha Ruiz de la Prada.
eLitter	Based on studying littering in the land environment with the aim of discovering its origin and finding ways of preventing it. This is a joint project with the *Asociación Paisaje Limpio*.
LessPack	Aims to encourage less packaging waste production and reuse.
INTEMARES- Caladeros Limpios (clean fishing grounds)	The main objective of this project is to establish general guidelines for fishing for marine litter. It is part of the LIFE INTEMARES project, coordinated by the Biodiversity Foundation from the Ministry for Ecological Transition and Demographic Challenge.
Mares circulares (circular seas)	The main objective of this project is to remove marine waste from the bottom of the sea, remove floating waste, manage the waste collected and reintroduce it into the production cycle. This project has been developed in partnership with other associations and with fishermen and fish markets in 16 ports in Spain and Portugal.
Marnoba	Its main objective is to collect, store and show information about marine waste in a simple way, via a mobile phone application (app MARNOBA), free to download and available for iOS and Android. It has been developed with support from the Biodiversity Foundation from the Ministry for the Ecological Transition and the Demographic Challenge (MITERD) and KAI Marine Services.
Oceanets	A project aimed at the prevention, recovery, re-use and recycling of fishing gear to produce added-value products in the textile industry. It has been devised by the *Instituto Tecnológico del Plástico* (AIMPLAS) in partnership with the University of Vigo, the Cooperative of Fishing Vessel Owners of the Port of Vigo, the Spanish sustainable fashion brand ECOALF and the Czech textile manufacturer SINTEX.
Libera Ciencia - Rios (liberate science - Rivers)	This project focuses on research into new systems to monitor and control pollution and microplastics in rivers.
RepescaPlas (re-fishing plastic)	Aims to reduce the impact of marine waste. It is managed by the Spanish Plastic Technological Centre (AIMPLAS) and is supported by various associations, public organisations and universities.

Source: Compiled by the Asociación Vertidos Cero (2022)

Table 3 *Recircular* project up to 2022

Project name
Recovery of leather offcuts from the automotive industry
Leftovers from the manufacture of tampons and panty liners
Secondary life for shampoo production by-products
Textile and wood post-consumption remains
Brewers grains and coffee grounds
Circularisation methodology for companies #Oleku
LessPack, strategies for reducing and reusing waste packaging
Circular innovation for band cards

Source: Compiled by Recircular (2022)

in order to develop a method of calculation based on a triple bottom line (economic, environmental and social).

Finally, any type of business, public administration, foundation, etc. can take part in the innovation projects developed by *Recircular*. Through these projects, *Recircular* can help them define and implement circular strategies and also implement innovation strategies to give a new life to materials that they no longer need and would otherwise be incinerated or thrown away. Table 3 shows some of the innovation projects developed by *Recircular*.

3 LessPack[4]

LessPack is a collaborative project between the three aforementioned organisations and was launched in 2019. Its main objectives are, firstly, to promote initiatives that reduce the generation of packaging waste and, secondly, to identify opportunities aligned with both the many legislative requirements and consumer demand for more sustainable consumption options.

In order to achieve these objectives, *LessPack* has undertaken various actions that can be divided into the following areas:

1. Strategies to reduce packaging waste
2. Training for businesses
3. A consumer perceptions study

3.1 Identification of Packaging Waste Reduction Strategies

The first phase of the *LessPack* project involved the identification of best practice in companies that have adopted measures to reduce their packaging waste. Following this analysis, the next phase involved selecting four reduction strategies that could be implemented in companies and businesses. They are as follows:

[4]The information in this sub-section was taken mostly from Miranda et al. (2021).

Fig. 2 Perfume dispensers. Source: Own figures

Strategy 1. Buy loose products using reusable packaging. Buying loose products using reusable packaging, whether manually, using dispensers and/or automatic or intelligent systems can reduce the amount of packaging used. To this end, the importance of using containers appropriate for maximum further reuse in the future under appropriate conditions should be highlighted. One example of best practice is the Spanish company *La Botica de los Perfumes* which offers consumers a wide range of perfumes and other products that can be bought loose or as refills (Fig. 2).

Strategy 2. Using solid and concentrated products. This involved the development of new products in solid or concentrated form which are usually available in liquid form (soaps, shampoo, etc.). Furthermore, these types of products are less bulky in nature, which also results in a reduction in waste packaging. Finally, the refills for solid and concentrated products come in smaller packaging, which also means less packaging is required. This strategy is being employed by an increasing number of companies in the cosmetics industry, offering consumers a wide range of solid soaps for facial, body and hair care (Fig. 3).

Strategy 3. Recovery of reusable waste packaging. There are two possible options available: consumers return brand-specific (monobrand) packaging using the postal service or companies collect packaging from different brands (multibrand) from consumers' homes, offering the possibility of waste packaging collection similar to online shopping deliveries. In addition to these two options, a reference example for the recovery of reusable packaging is the Return and Reward Schemes in different countries. One of the ways in which these schemes are delivered is through specially designed machines, like a vending machine but in

Fig. 3 Solid soap. Source: Royalty-free image from pixabay.com

Fig. 4 Collection machine for reusable packaging. Source: Royalty-free image from pixabay.com

reverse, which uses a mechanism that enables the collection of reusable packaging (Fig. 4). The main purpose of the machines is to encourage recycling outside the home through schemes offering prize draws for electric scooters, sustainable mobility initiatives and donations to social or environmental projects, etc. The machines can be found in public spaces, for example, transport hubs, shopping centres, hospitals and universities.

Strategy 4. Material innovation in packaging manufacturing. In this area, materials are made from renewable raw materials of biological origin, materials obtained from waste and by-products from different manufacturing processes and finally recycled and recyclable materials. Using these raw materials and other materials, an increasing number of final products can be manufactured, such as those relating to fashion (Fig. 5). An example of good practice in this area is the Spanish sustainable fashion brand ECOALF, whose products are produced using waste extracted from the oceans. Furthermore, a similar strategy has also

Fig. 5 Sustainable fashion. Source: Royalty-free image from pixabay.com

been developed by international brands such as Adidas and Nike in the production of some of their products.

3.2 Training and Advice for Businesses

In 2020, the project delivered a series of sessions to present *LessPack* to manufacturers and businesses, explaining the packaging waste reduction initiatives being developed, as well as providing information on the new regulatory framework for waste packaging expected to be approved in the European Union.

Since its founding in 2021, *LessPack* has been offering a call for proposals from businesses (both manufacturers and distributors) that want to implement initiatives to reduce the volume of the waste packaging they produce. The aim was to support these businesses throughout the process, adapting to each particular case and treating them as pilot projects.

3.3 Consumer Perception Study

The effectiveness of the design changes to packaging relies heavily on final consumer acceptance. In view of this, in 2020, *LessPack* carried out a consumer perception study to find out how waste packaging reduction and disposal initiatives being developed were being received.

Study participants included representatives from key Spanish consumer, neighbourhood and citizens' organisations. They were invited to give their opinions on the motives and barriers which exist for consumers in adopting different waste packaging reduction and disposal measures identified in the *LessPack* project.

By way of example, Table 4 shows the main motives and barriers to participation in the initiatives developed by *LessPack* identified in this study.

Finally, it is worth mentioning the recognition received by *LessPack* from the "#PorElClima" community, in the fourth edition of the "101 Ejemplos Empresariales #PorElClima" celebrated in 2021. This edition recognises climate

Table 4 Key motives and barriers to *LessPack* initiatives

Key motives	Key barriers
• Personal satisfaction from contributing to the benefit of all • Personal benefit • Reducing the amount of packing • New and interesting initiatives	• Doubts over the benefits of the initiatives • Packaging security, in particular relating to food • Change in habits and in planning • Lack of information • Price being a limiting factor to purchasing

Source: Compiled by Miranda et al. (2021)

commitment and ambition in the business world, with the aim of identifying specific initiatives addressing the climate emergency and supporting decarbonisation action (Comunidad #PorElClima, 2021).

4 Conclusions

Packaging plays a fundamental role in modern trading systems. It not only contains the product but also enables the logistics process, including transport, warehousing and display on retail shelves. Furthermore, it is true to say that packaging is the "silent salesperson" of the product, because it attracts consumers' attention on retail shelves full of very similar products. For this reason, many companies opt for overpackaging, using excessive packaging and designs which require multiple materials.

The packaging itself has a negative side, due to its detrimental impact on the environment. Firstly, as natural resources are used in its manufacture, and secondly, due to the amount of waste generated once, the product has been used or consumed.

Circular economy is the answer. This involves changing production and consumption models to use natural resources more efficiently and extend the useful life of products and their different components and materials. Recycling and reuse are two ways to achieve this. Nonetheless, priority must be given to preventing contamination, in other words, avoiding waste generation. In view of this, many circular economy strategies should be based on reduction. How can packaging waste be reduced? One solution is to redesign packaging, where contaminating materials (such as plastic) are replaced by biodegradable or sustainably sourced ones. Another solution is to eliminate all or some of the packaging used to market certain types of products. The most striking example is replacing packaged products with loose products and refills.

The *LessPack* project is a collaboration between three organisations that aim to provide different circular economy options to manufacturers, retailers and consumers that can be applied to packaged products. Beyond the legal requirements being developed in many countries to reduce packaging, success in reducing packaging waste will depend on consumer attitudes when it comes to changing their habits and patterns in favour of more ecological packaging or even package-free buying.

5 Discussion Questions

Question 1: Can you find out information about initiatives adopted by manufacturers or businesses to reduce, recycle or reuse packaging waste or use ecological designs? Describe the packaging initiative or improvement. Critically reflect on the positive impact it could have.

Question 2: What packaging waste does your household produce? List the types of packaging waste you produce over several days, type of product used for and type of material (cardboard, plastic, glass, etc.), and then critically reflect on their use and how they could be ecologically redesigned.

Question 3: What are your shopping patterns relating to the product and your disposition to recycle the waste packaging, reuse it and even to buy non-packaged products? Reflect on your reasons for taking part in these initiatives, any barriers to participation and how these barriers could be overcome and your general opinion on the initiatives.

Acknowledgements The authors acknowledge the financial support provided by the Junta de Extremadura and FEDER under the grant GR21078.

Appendix: Teaching Notes for the Case

Synopsis

One of the biggest threats to the sustainability of the planet is the growing volume of waste produced as a result of current production and consumption models. According to Eurostat, the volume of waste generated per European inhabitant has gradually increased in the last decade.

In response to this reality, various international organisations such as the UN and the European Union have established different action plans focusing on the circular economy. The aim is to avoid waste and encourage waste processing to produce high-quality secondary resources from it, which can then be integrated into an effective secondary raw materials market.

In 2019, the *LessPack* initiative was founded in Spain to support businesses, distributors and consumers. This collaborative project includes the involvement of non-profit organisations Ecoembes and the *Asociación Vertidos Cero*, as well as the private entity *Recircular*. Despite differences in their legal status, all of them are united in their concerns about the negative impact waste has on the planet.

The project promotes a reduction in packaging waste production and packaging reuse. The actions undertaken by this project can be divided into three main areas: identifying strategies to reduce packaging waste, delivering training activities and advice to businesses and developing a study on the consumer perceptions on this issue.

Potential Audience and Instructor's Material

This case has been developed for use in the framework of socially responsible consumption and in particular the reduction of the generation of packaging and its reuse.

The potential audience for this chapter is twofold:

- Undergraduate students pursuing degrees in business administration or marketing
- Graduate students pursuing MBA and masters in marketing or executive programs and seminars (on socially responsible consumption in general and the reduction of packaging generation and its reuse in particular)

Learning Objectives

- Reflect on the impact generated by packaging waste on planet sustainability.
- Understand the importance of innovation and eco-design in new packaging to prevent waste generation and to facilitate the recycling and reuse of packaging already produced.
- Learn about the key motives and barriers to participation in the initiative.

Time Frame for Class Discussion

This case can be taught in one session of 60–100 minutes, depending on the number of students participating in the discussion activities, as well as the previous work that students have done before class.

Suggested Discussion Questions

1. Can you find out information about initiatives adopted by manufacturers or businesses to reduce, recycle or reuse packaging waste or use ecological designs? Describe the packaging initiative or improvement. Critically reflect on the positive impact it could have.
2. What packaging waste does your household produce? List the types of packaging waste you produce over several days, type of product used for and type of material (cardboard, plastic, glass, etc.), and then critically reflect on their use and how they could be ecologically redesigned.
3. What are your shopping patterns relating to the product and your disposition to recycle the waste packaging, reuse it and even to buy non-packaged products?

Reflect on your reasons for taking part in these initiatives, any barriers to participation and how these barriers could be overcome and your general opinion on the initiatives.

Analysis

Answer to Question 1: *Can you find out information about initiatives adopted by manufacturers or businesses to reduce, recycle or reuse packaging waste or use ecological designs? Describe the packaging initiative or improvement. Critically reflect on the positive impact it could have.*

This is a classroom-based group activity. Students will be divided into groups of four.

Each group will find out information about an initiative adopted by a manufacturer or business to reduce, recycle or reuse packaging waste or use ecological designs. Students will be asked to answer the following questions:

- Initiative name
- Explain what the initiative is: objective (preventing waste generation, reducing waste, recycling waste, reusing waste), actions undertaken, etc.
- Critically reflect on the positive impact it could have.

A table similar to the one below can be used for information gathering:

Initiative name	Initiative objective	Initiative actions	Reflection on the positive impact of the initiative

When the task has been completed, the members of each group will share ideas and create a presentation. Each group will present the initiative studied to the rest of the students.

Answer to Question 2: *What packaging waste does your household produce? List the types of packaging waste you produce over several days, type of product used for and type of material (cardboard, plastic, glass, etc.), and then critically reflect on their use and how they could be ecologically redesigned.*

This is a classroom-based group activity. Students will be divided into groups of four.

Each group should identify the packaging waste produced in their own households over several days and critically reflect on it. To do this, they must record the following information:

- Type of product the waste is produced from: juice carton, tinned food, highly processed foods, etc.
- Type of waste material: cardboard, plastic, glass, etc.

- Critically reflect on the usefulness of the packaging waste, if it is needed and how it could be ecologically redesigned

A table similar to the one below can be used for information gathering:

Type of product waste produced from	Type of waste material	Usefulness of the waste	Need for packaging waste	How it could be ecologically redesigned

When the task has been completed, the members of each group will share ideas and create a presentation. Each group will present the initiative studied to the rest of the students.

Answer to Question 3: *What are your shopping patterns relating to the product and your disposition to recycle the waste packaging, reuse it and even to buy non-packaged products? Reflect on your reasons for taking part in these initiatives, any barriers to participation and how these barriers could be overcome and your general opinion on the initiatives.*

This is a classroom-based group activity. Students will be divided into groups of four.

Each group should discuss the following questions:

- Motives for participation in these types of initiatives: caring for the environment, personal benefit, etc.
- Barriers to participating in these types of initiatives: time involved in recycling, reuse or buying loose products, cost of reusable containers, etc.
- How the identified barriers can be overcome.
- General opinion on these types of initiatives.

A table similar to the one below can be used for information gathering:

Motives for participation in these types of initiatives	
Barriers to participation in these types of initiatives	
How the identified barriers can be overcome	
General opinion on these types of initiatives	

When the task has been completed, the members of each group will share ideas and create a presentation. Each group will present the initiative studied to the rest of the students.

References

Asociación Vertidos Cero. (2022). *Official information from the website*. Consulted July 29, 2022, from https://vertidoscero.com/.

Comunidad #PorElClima. (2021). *Informe #101EjemplosPorElClima2021*. https://vertidoscero.com/lesspack_premio/

Ecoembes. (2022). *Official information from the website*. Consulted July 29, 2022, from https://www.ecoembes.com/.

European Commission. (2020, March 11). Modificar nuestras pautas de producción y consumo: El nuevo Plan de acción para la economía circular muestra el camino hacia una economía competitiva y climáticamente neutra de consumidores empoderados. Press Release.

Eurostat. (2022). *Municipal waste generation*. Consulted July 31, 2022, from https://ec.europa.eu/eurostat/

Miranda, F. J., Astrain, P., Villaluenga, N., & López-Samaniego, E. (2021, 31 May to 3 June). *LessPack. Estrategias para la reducción de los residuos de envases* [Written document]. 15° Congreso Nacional de Medio Ambiente (CONAMA) 2020, Centro de Convenciones Norte/Institución Ferial de Madrid (IFEMA), Madrid, Spain.

Recircular. (2022). *Official information from the website*. Consulted July 29, 2022, from https://recircular.net/.

Corporate Social Responsibility of "Ant Forest": Ant Group's "Internet + Tree Planting" Public Welfare Project

Estela Núñez-Barriopedro, Pedro Cuesta-Valiño, and Yanhua Yang

Learning Objectives

1. Understand the important role of corporate social responsibility in achieving economic growth and sustainable development.
2. Based on the value co-creation theory, this paper analyzes the advantages of Ant Forest project and how to reflect corporate social responsibility.
3. Understand how all relevant parties in the Ant Forest project are connected and how to achieve mutual benefit and win-win results.
4. Study the operation modes of various public welfare projects in the world that use the Internet as a platform.

1 Introduction

Ant Forest is a public welfare project launched by Alipay on the Internet, aiming to drive the public to reduce carbon emissions. The Ant Forest project is currently mainly running in China. The low-carbon behavior of the participants of the Ant Forest project can be counted as "green energy" in the Ant Forest. When the "green energy" has accumulated to a certain level, the participants can use a mobile phone to apply for planting a real tree in an area in urgent need of ecological restoration or to "claim" protection rights in an area in which biodiversity is in urgent need of protection (Zhang et al., 2020).

According to "Ant Group 2021 Sustainability Report" (Ant Group, 2022), Ant Forest has integrated 43 low-carbon use cases and facilitated more than 600 million people making low-carbon life choices, and users generated more than 20 million tons of "green energy" points through making low-carbon life choices. Total trees

E. Núñez-Barriopedro (✉) · P. Cuesta-Valiño · Y. Yang
Universidad de Alcalá, Alcalá de Henares, Madrid, Spain
e-mail: estela.nunezb@uah.es; pedro.cuesta@uah.es; yanhua.yang@edu.uah.es

E. C. Bianchi et al. (eds.), *Responsible Consumption and Sustainability*, Springer Business Cases, https://doi.org/10.1007/978-3-031-30742-3_7

planted with funding from donations made by Ant Group reached 326 million across 3.97 million mu (approx. 264,667 ha.) of land.

This project has promoted green and low carbon to become a popular trend in China. At the same time, this model is also used in more countries. This model is also borrowed by GCash, a mobile payment platform in the Philippines. With the technical support of Alipay, a subsidiary of Ant Financial, the "GCash Forest" launched in June 2019 encourages Filipinos to participate in environmental protection and afforestation.

In 2019, Ant Forest won the United Nations' highest award in the field of environmental protection, the "Champions of the Earth," and the highest award for addressing climate change, the "Lighthouse Award."

As a social-type public welfare project, Ant Forest project innovatively adopts the "public welfare" and "gamification" model, considering public welfare value and market economic value. The project organizer and participants are on a platform built by the Internet. The carbon emissions saved by participants in low-carbon travel, online payment, ticket consumption, etc. will be calculated as virtual "green energy." After accumulating a certain amount of "green energy," participants can apply for planting trees in Ant Forest. Participants obtain green energy through low-carbon behaviors, including green travel, travel reduction, high efficiency and energy saving, recycling, paper reduction, and plastic reduction (see Table 1).

After participants use low-carbon behaviors, they can accumulate "green energy" in their "personal carbon account" the next day. At the same time, on the Ant Forest platform, participants can become "friends" with another participant. After becoming a "friend," participants can water their friends' trees, steal their friends' "green energy," use a protective cover to protect the "green energy," use accelerators to accelerate the maturity of "green energy," check the "green energy" ranking list, etc. Ant Forest has rich gamified operations. It reflects the social attributes of Ant Forest.

In addition to the single-player mode, Ant Forest has also launched a combined planting mode, including love tree, family tree, friend tree, classmate forest, colleague forest, and pioneer forest. In the co-planting mode, participants invite family members, friends, etc. to collect green energy together to speed up the process of planting trees. In the process of co-planting, the interaction between the participants is enhanced, and the emotional pursuit of the participants is stimulated. This is a further manifestation of the social nature of Ant Forest.

When the green capacity in the participant's "personal carbon account" accumulates to a certain amount, the participant can exchange the right to plant according to their own preferences, including the type of plant, the location of planting, and even a meaningful selection for this plant's name (Table 2).

Table 1 The method of Ant Forest to obtain green energy[a]

Type	Low-carbon behavior	Green energy available
Green travel	Go on foot	Convert according to the actual situation
	Bike-sharing	Up to 159 g
	Bus travel	80 g per transaction
	Subway travel	52 g per transaction
	New energy vehicle	891 g per day
	Public electric vehicle charging station	Up to 960 g per transaction
Travel reduction	Buy train tickets online	136 g per transaction
	Online ticket purchase	180 g per transaction
	Green government affairs	15 g per transaction
	Living expenses	262 g per transaction
	Green medical	Up to 15 g
	Green office	Up to 51 g per day
	Less drive	819 g per week
	Green bank	Up to 639 g
	Credit card payments	21 g per transaction
High efficiency and energy saving	ETC (Electronic Tool Collection) payment	23 g per transaction
	Sharable chargers	13 g per transaction
	Pay for parking	18 g per transaction
	Green appliances	18,400 g per transaction
Recycling	Green recycling	Convert according to the actual situation
	Green stay	92 g per stay
Paper reduction and plastic reduction	Green packaging	Convert according to the actual situation
	Offline payments	5 g per transaction
	Electronic receipt	4 g per transaction
	Electronic invoice	5 g per transaction
	Green takeaway	16 g per transaction
	Environmentally friendly plastic reduction	12 g per transaction
	Green cup	Up to 600 g per transaction
	Direct drinking water	4 g per transaction
	Scan code to order	7 g per transaction
	Scan the QR code to buy tickets	5 g per transaction
	Online shipping	4 g per transaction
	Electronic contract	6 g per transaction
	Electronic insurance policy	59 g per insurance policy
	Credit live	5 g per transaction
	International tax refund	4 g per transaction
	Paperless reading	Up to 150 g per day

Source: Own elaboration based on Ant Forest (2022)
[a]Green energy is provided by the Beijing Environmental Exchange, The Nature Conservancy, etc. to provide carbon emission reduction and carbon sink scientific algorithms

Table 2 Green energy required for different types of trees

Number	Tree category	Green energy needed
1	Apricot tree	38,570 g
2	Sacsaoul	17,900 g
3	Caragana	16,930 g
4	*Hedysarum scoparium*	21,310 g
5	*Salix psammophila*	19,680 g
6	Sea buckthorn	18,880 g
7	Chinese tamarisk	22,400 g
8	Mountain peach	37,980 g
9	Pinus sylvestris	146,210 g
10	Arborvitae	96,000 g
11	Chinese pine	114,000 g
12	China Armand pine	185,000 g
13	Spruce	198,000 g
14	Diversifolius poplar	215,680 g

Source: Own elaboration based on Ant Forest (2022)

2 Case Development

2.1 Case Background

Responsible consumption is defined as "the consumption that has less negative impact or more positive impact on the environment, society, the self, and the other-beings." In the market, supply and consumption are mutually reinforcing and inseparable. With more and more green consumption options available, personal green low-carbon consumption will promote the transformation and upgrading of technology and R&D on the production side, thus promoting the transformation and upgrading of the whole market and social economy (Núñez-Barriopedro and Llombart Tárrega, 2021). If an enterprise wants to sustain economic growth, it must consider its impact on the environment and society, and its social responsibility is particularly important.

This case introduces the Ant Forest project created by Ant Group. Ant Forest aspire to promote a green lifestyle among the public through technology and model innovation to empower environmental protection and sustainable socioeconomic development.

From the United Nations website, [1] droughts are among the greatest threats to sustainable development, especially in developing countries but increasingly so in developed nations too. In fact, forecasts estimate that by 2050, droughts may affect over three quarters of the world's population. The number and duration of droughts has increased by 29% since 2000, as compared to the two previous decades. When

[1] World Day to Combat Desertification and Drought, 17 June: https://www.un.org/en/observances/desertification-day

more than 2.3 billion people already face water stress, this is a huge problem. More and more of us will be living in areas with extreme water shortages, including an estimated one in four children by 2040. No country is immune to drought. This year, the theme of the International Day Against Desertification and Drought is "Rising up from drought together." Spain is the Desertification and Drought Day 2022 host. The theme emphasized the need of an early action to avoid disastrous consequences for humanity and the planetary ecosystems.

It can be seen from the above that the desertification problem has become an important problem that restricts human development and threatens human existence. It is urgent to get attention from all walks of life, and more energy and resources are needed to solve the problem of desertification. The development of Internet technology has put forward new ideas and possibilities for solving the problem of desertification.

Following ESG (environmental, social, and governance) guidelines, Ant Group prioritizes sustainability and integrates it into the company's overall development strategy. Ant Forest users can generate green energy from low-carbon deeds to redeem real trees. Ant Forest cooperates with public welfare institutions to plant these real saplings in desertification areas, so as to improve the environment. The "personal carbon account" function of Ant Forest users has enabled the organic integration of the Internet and green finance. It provides new ideas and explores new paths for financial participation in environmental governance. Ant Forest has become a window to encourage the public to understand and participate in green low carbon.

2.2 Analysis of Ant Forest Project Based on Value Co-creation Theory

2.2.1 Value Co-creation Theory

In the past, due to the asymmetry of information, the traditional theory of value creation believed that enterprises or producers were the only value creators, and consumers were simply using value (Ravina-Ripoll et al., 2021). Now, consumers have more and more ways to obtain information, and consumers' ability to use information is getting stronger and stronger (Alonso-Garcia et al., 2022). Consumers are also involved in the creation of value. Businesses or producers are no longer the only value creators (Núñez-Barriopedro et al., 2020). Consumer behavior plays an increasingly important role in value creation (Cuesta-Valiño et al., 2022).

Normann and Ramírez (1993) put forward the idea of value co-creation, arguing that the interaction between suppliers and consumers is an important part of value creation. In this case, value co-creation means that each stakeholder in Ant Forest's public welfare social network aims to increase the value of the entire tree planting activity, trust each other, cooperate with each other, share resources, and at the same time realize the growth of their own interests.

Specifically, in Ant Forest's "Internet + tree planting" public welfare social case, value co-creation is of great significance to both the Ant Forest platform and stakeholders. By involving stakeholders in value co-creation, it helps Ant Forest to

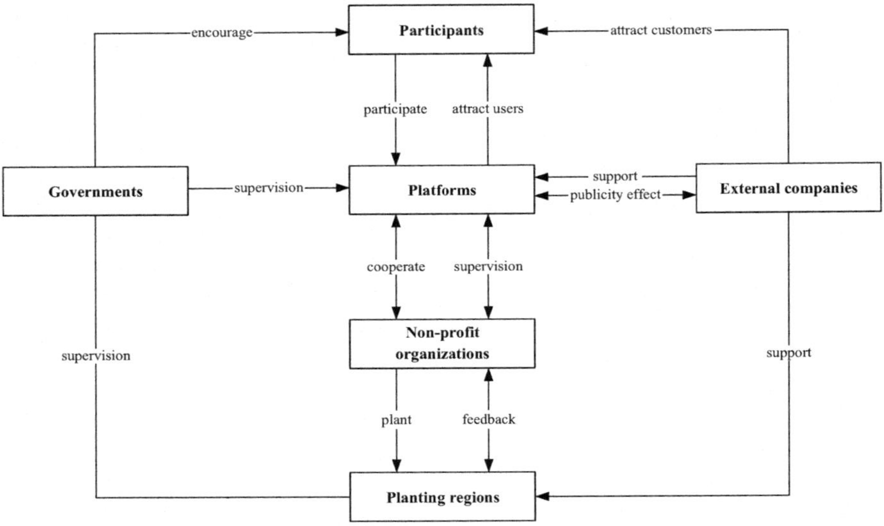

Fig. 1 Interrelationships among various stakeholders in Ant Forest. Source: Own elaboration based on Ant Forest (2022)

improve service quality, reduce costs, improve efficiency, discover market opportunities, invent new products, improve existing products, increase brand awareness, and enhance brand value, competitor's competitive advantage, etc. (Núñez-Barriopedro et al., 2021). By participating in value co-creation, participants can get a sense of achievement or reward in low-carbon activities and obtain unique experience through the interaction of the whole value co-creation. These gains of participants have further impact on Ant Forest, such as improving participants' satisfaction, loyalty, willingness to continue to participate, etc.

2.2.2 How to Contact Stakeholders in the Ant Forest Project

In the past, the public welfare communication campaign of afforestation on the Internet was only a single content communication strategy. The Internet only played a role of information transmission and was only an information communication platform. As a social-type public welfare project, Ant Forest innovatively forms an action platform on the Internet that everyone can participate in.

In the Ant Forest project, there are six stakeholders: participants, platforms, nonprofit organizations, planting areas, governments, and external companies. There are connections among them (see Fig. 1).

The roles and behaviors of six stakeholders in value co-creation constitute the success of Ant Forest. Thanks to the development of network information technology, public welfare is no longer a complicated and vast project, and public welfare has become more convenient and simple. With the help of the network platform, the conversion of virtual products and real currencies is set up, which reduces the public welfare cost of participants, improves corporate social responsibility, and achieves a win-win situation for all parties. The six stakeholder roles are described below:

1. Participants refer to the people who participate in the Ant Forest project. As long as you register an account on Alipay, you can participate in the Ant Forest project.
2. Platform refers to Alipay, the Internet platform where the Ant Forest project is located. Alipay is a product of Ant Group. Alipay (China) Network Technology Co., Ltd., established in 2004, is a third-party payment platform in China.
3. Nonprofit organizations refer to the nonprofit organizations that cooperate with the Ant Forest project, including the China Green Foundation, Alxa SEE Foundation, Alxa Ecological Foundation, etc. The Ant Forest project will entrust nonprofit organizations to plant in designated locations. These nonprofit organizations also supervise planting projects.
4. Planting area refers to the main beneficiaries of the Ant Forest project, including ecologically fragile areas such as deserts and plateaus.
5. The government refers to the government of the country in which it is located. On the one hand, the government plays a role in supervising the project. On the other hand, the government also actively advocates citizens to participate in low-carbon projects such as "Internet + tree planting" and will issue a "Certificate of Responsibility for National Obligation to Plant Trees" to participants who have achieved success.
6. External companies refer to the companies that sponsor the Ant Forest project. For example, on the "National Low-Carbon Day" on August 25, 2021, the first batch of more than 100 companies including Haier, FAW-Volkswagen, and Muji participated in Ant Forest's "Green Energy Action." Together, Ant Forest's "green energy" is used as an integral reward to promote consumers to choose more energy-saving, low-carbon, and emission-reducing products and services, so as to promote the whole society to form a moderately economical, green, and low-carbon lifestyle and environmental protection fashion.

   The Ant Forest project makes good use of relational chains. Especially in the era of increasingly socialized Internet, Ant Forest closely connects six stakeholders. In this way, all stakeholders have realized value co-creation and fully reflected the corporate social responsibility.

2.2.3 The Value Co-creation and Development Process of Stakeholders in the Ant Forest Project

The value co-creation development process of all parties in the Ant Forest project can be described as "value co-creation demand—value co-creation action—value co-creation result." In the process of value co-creation, participants, platforms, nonprofit organizations, planting regions, governments, and external companies all have their own needs. Value co-creation actions were taken driven by demand. Finally, the result of value co-creation is achieved: to achieve a mutually beneficial and win-win situation. This is also the effect that Ant Forest's "Internet + tree planting" public welfare project wants to achieve.

The value co-creation process can be analyzed using Table 3.

Table 3 Co-creation and development process

	Value co-creation demand	Value co-creation action	Value co-creation result
Participants	1. The need for participation in environmental protection 2. Incentives for Ant Forest gamification design 3. Social needs (Ant Forest has social attributes)	1. Registration 2. Participate 3. Introduce more users to participate	Participants gain: the sense of achievement of public welfare participation, the emotional energy brought by public welfare results, the fun and spiritual motivation brought by the gamification participation mechanism, and the sense of social experience
Platforms	1. Ant Group aims to create social change and realize a low-carbon life through the Ant Forest project. 2. Attract more users to use Alipay through the development of the Ant Forest project and increase user stickiness	Pay attention to ecologically fragile areas such as deserts and plateaus, and donate funds to public welfare organizations	Attract more new users to use Alipay and increase the loyalty of old users
Nonprofit organizations	Achieving the Sustainable Development Goals (SDGs), the need for environmental protection	Organize specific work such as planting and maintenance	Environmental sustainability
Planting regions	Environmental protection needs	Actively cooperate with planting, maintenance, and other works	Conserve soil and water and prevent desert expansion
Governments	1. Environmental protection goals and governance tasks 2. Stimulate the enthusiasm of the public to participate in environmental protection	The local forestry department conducts business supervision	Achieving environmental protection goals and governance tasks
External companies	1. Corporate social responsibility 2. To enhance brand image through public welfare marketing strategy 3. Improve the economic benefits of the enterprise	1. Participate in the "green energy" plan when selling products 2. Donation	Directly enhance the company's brand image. Indirectly improve the economic efficiency of enterprises

Source: The above data comes from own elaboration based on Ant Forest (2022)

Ant Forest uses commercial marketing methods to achieve social welfare purposes. On the other hand, Ant Forest also uses social welfare value to promote its own business services. While continuing to meet the needs of the six stakeholders, Ant Forest itself has also acquired loyal users and increased user stickiness. From a long-term perspective, a more reasonable consideration of the immediate and long-term interests of all stakeholders has contributed to the realization of the sustainable development goals.

Relying on the "digital life platform" formed by Alipay, it is expected that companies participating in the "Green Energy Action" will cover many aspects of daily life such as "clothing, food, housing, transportation, and use," which will affect hundreds of millions of consumers. More and more companies with a sense of social responsibility have joined Ant Forest, supporting "visible green" with corporate donations, inspiring the public to choose a green and low-carbon lifestyle, and creating "invisible green" with practical actions. Reshape low-carbon environmental protection and green consumption behavior under the digital transformation of economy and society.

3 Conclusions

Ant Forest is a successful case of Internet public welfare CSR project.

Ant Forest attracts participants with unique and interesting creative content. Participants are eager to share information with others, so as to gain recognition and praise from others, and intend to satisfy their spiritual needs through the recognition of more audiences. The Ant Forest reflects its unique characteristics in the way of transmission. The communication content of Ant Forest is virtual green energy, which forms a closed loop of production and sales around energy and realizes public welfare communication. Socialize through green energy interactions. This case has a wide range of audiences, forming a social atmosphere of "everyone participates in public welfare."

From the perspective of value co-creation, in the Ant Forest project, participants, platforms, nonprofit organizations, planting areas, governments, and external companies are all interconnected. During this process, all parties have well perfected the role-playing and actively promoted the publicity of the Ant Forest activity.

This model also provides sustainable support for public welfare tree planting and low-carbon emission reduction in various countries.

4 Discussion Questions

1. Ant Forest uses the "Internet + tree planting" public welfare method. What are the advantages?
2. For Ant Forest's "Internet + tree planting" public welfare project, what do you think can be improved?

3. Does your country have corresponding measures to encourage people to participate in planting trees and protecting forests? If so, are these measures currently effective? How enthusiastic is the public to participate?

Appendix: Teaching Notes for the Case

Synopsis

Alipay Network Technology Co., Ltd., is a third-party payment platform in China. Ant Forest is a public welfare project of Alipay on the Internet that aims to drive the public to reduce carbon emissions. The Ant Forest project was created in 2016. The low-carbon behavior of participants can be transformed into "green energy" on the Ant Forest platform. The "green energy" accumulated to a certain level can be exchanged for the planting of real trees. Ant Forest's planting projects and ecological restoration projects in various places are donated by Ant Group to nonprofit organizations, which organize specific work such as planting and maintenance, and are supervised by the local forestry department. All projects have corresponding donation agreements and acceptance reports. Ant Forest uses the idea of value co-creation to realize the "Internet + tree planting" public welfare model. The Ant Forest project won the United Nations "Champions of the Earth Award" in 2019.

Potential Audience and Instructor's Material

This case can be used as study material for marketing and marketing management courses. Because this chapter takes into account social development, business models, etc., it can also be used in several other courses, such as economics, sociology, regional development, public sector marketing, and marketing strategies, among others.

Learning Objectives

1. Understand the important role of corporate social responsibility in achieving economic growth and sustainable development.
2. Based on the value co-creation theory, this paper analyzes the advantages of Ant Forest project and how to reflect corporate social responsibility.
3. Understand how all relevant parties in the Ant Forest project are connected and how to achieve mutual benefit and win-win results.
4. Study the operation modes of various public welfare projects in the world that use the Internet as a platform.

Time Frame for Class Discussion

This case can be taught one session of 45 minutes approximately.

Suggested Discussion Questions

1. Ant Forest uses the "Internet + tree planting" public welfare method. What are the advantages?
2. For Ant Forest's "Internet + tree planting" public welfare project, what do you think can be improved?
3. Does your country have corresponding measures to encourage people to participate in planting trees and protecting forests? If so, are these measures currently effective? How enthusiastic is the public to participate?

Analysis

Answer to Question 1: *Ant Forest uses the "Internet + tree planting" public welfare method. What are the advantages?*

Advantage 1: In the Ant Forest project, six stakeholders are closely linked to achieve value co-creation. Stakeholders include participants, platforms, nonprofit organizations, planting regions, governments, and external companies. There are connections between the various stakeholders involved. The value co-creation development process of all stakeholders in the Ant Forest project can be described as "value co-creation demand—value co-creation action—value co-creation result." All six stakeholders have realized their own needs and ultimately contributed to the realization of the SDGs.

Advantage 2: From the perspective of corporate social responsibility (CSR), it is more reasonable to take into account the immediate interests and long-term interests of the six stakeholders Galán-Ladero (2012). The Ant Forest project pursues long-term and sustainable development.

Advantage 3: The interactive way of Ant Forest. The CSR of Ant Forest has formed good interactive feedback, which enriches the user's sense of social experience. Participants shared the sense of achievement of public welfare participation, the emotional energy brought by public welfare results, and the fun and spiritual motivation brought by the gamification participation mechanism, thus forming the recognition of public welfare, forming "interaction—continuous participation—continuous participation—interaction" virtuous circle. This model provides sustainable support for public welfare tree planting and low-carbon emission reduction.

Answer to Question 2: *For Ant Forest's "Internet + tree planting" public welfare project, what do you think can be improved?*

With the continuous development of the mobile Internet, the dissemination scope and dissemination influence of the Ant Forest project are gradually increasing.

However, the Ant Forest project still has shortcomings. We discuss some deficiencies and future development countermeasures here:

1. The planting cycle of the Ant Forest project is long, which weakens the enthusiasm of the participants. Ant Forest contains a total of 14 kinds of saplings, only a small number of saplings can mature without too much green energy, which can quickly meet users' demands for public welfare, growing big trees and achieving results. However, saplings such as Diversifolius poplar and Spruce require more than 190,000 g of green energy, and it takes a long time to accumulate the required energy. During the long accumulation period, many impatient users gradually faded out of this public welfare activity.
   Therefore, it is recommended that the Ant Forest project launch more new functions and upgrade services to improve the user's participation experience. Internet public welfare should continuously update personalized experience and interesting elements, so that users can participate in the construction of public welfare communication while enjoying the fun of the game (Pan, 2020).
2. The Ant Forest project needs to strengthen the long-term interactive mechanism. The Ant Forest project mainly relies on the social relationship chain to spread and spreads through the mutual sharing of participants. Therefore, it is recommended that the Ant Forest project set up related interactive topics on social platforms and conduct long-term interaction and sharing around the topics (Pan, 2020). Suggest an interaction mechanism that can maintain participants' emotions in the long term (Núñez-Barriopedro et al., 2021). It helps to consolidate the short-term emotional experience of participants into stable emotional energy, thus forming a stable power source for public welfare communication (Penelas-Leguía et al., 2012; Núñez-Barriopedro, 2019).
3. Ant Forest needs to provide timely feedback on offline tree planting results in more ways. Environmental public welfare is a project with a long duration and a short-term effect that is difficult to measure. Controlling desertification by planting trees is also a complex and long-term project. Therefore, the Ant Forest project needs to timely feedback the situation of offline tree planting. Let online participants get real feedback on offline tree planting faster. The methods that have been taken so far: real-time feedback of the real scene ahead, forest rangers telling stories about sticking to the forest, drawing online participants to visit the site, etc. However, there are still many online participants who are unable to obtain the information they want in a timely manner. Therefore, the Ant Forest project needs to enhance and improve the participants' public welfare participation experience in terms of content, disclosure mechanism, and online and offline linkages and strengthen the participants' identification and trust in public welfare groups (Pan, 2020).

Answer to Question 3: *Does your country have corresponding measures to encourage people to participate in planting trees and protecting forests? If so, are these measures currently effective? How enthusiastic is the public to participate?*

1. *Arbor Day in various countries*

All countries in the world attach great importance to tree planting and protecting the ecological environment. At present, more than 50 countries have established Arbor Days. Due to the different national conditions and geographical locations of different countries, the name and time of Arbor Day are also different in different countries.

The details of Arbor Day in some countries are as follows.

	Country	Arbor Day
1	China	Chinese Arbor Day is March 12th.
2	Spanish	The Spanish Arbor Day is February first, and the City of Madrid regards the first week of February first as "Tree Planting Week" every year.
3	UK	Arbor Day in the UK is November 6–12.
4	France	March 31 is Tree Day and March is Greening Month.
5	Senegal	The Senegal Arbor Day is from early August to mid-October, and national tree planting activities begin after the rainy season every year.
6	Greece	Arbor Day in Greece is fall.
7	Italy	Arbor Day in Italy is November 21st.
8	Irish	Irish Arbor Day is March 17th.
9	Pakistan	Arbor Day in Pakistan is August 4th.
10	Egypt	Arbor Day in Egypt is from September to November.

2. *Activities about planting trees*

Some countries or organizations will also initiate activities related to tree planting, for example, the United Nations Strategic Plan for Forests 2017–2030. [2] "The agreement on the first-ever UN Strategic Plan for Forests was forged at a special session of the UN Forum on Forests held in January 2017 and provides an ambitious vision for global forests in 2030. The plan was adopted by the UN Economic and Social Council on 20 April 2017 and was subsequently adopted by the UN General Assembly on 27 April 2017. The Strategic Plan features a set of six Global Forest Goals and 26 associated targets to be reached by 2030, which are voluntary and universal. It includes a target to increase forest area by 3% worldwide by 2030, signifying an increase of 120 million hectares, an area over twice the size of France."

For example, in 2020, US president Trump recently announced that the United States will join the Trillion Tree Campaign, a movement to plant, restore, and protect the number of trees around the world.

For example, in commemoration of the 70th anniversary of Queen Elizabeth II's accession to the throne, the UK launched the "The Queen's Green Canopy" tree

[2] The following information was obtained from the United Nations website: https://www.un.org/esa/forests/documents/un-strategic-plan-for-forests-2030/index.html.

planting activity, encouraged people to plant saplings, and planned to protect 70 woodlands.

For tree planting activities in more countries, students can check on the Internet.

3. *Tree planting achievement*

Tree planting activities in various countries have also achieved great results. Public participation is very enthusiastic. For example, in China, the government continues to promote afforestation projects and calls on the public to join. Data from the State Forestry and Grassland Administration of China shows that since 2012, a total of 150 million acres of afforestation have been completed. The forest coverage rate increased by 2.68 percentage points to 23.04%; the net increase of forest stock was 3.839 billion m^3, reaching 17.56 billion m^3; and the total carbon storage of forest vegetation increased by 1.375 billion tons, reaching 9.2 billion tons.

Many tree planting public welfare activities involve enterprises. At the same time, the enterprise takes into account the needs of achieving economic growth and sustainable development, encourages the public to participate in green and low-carbon activities, and reflects the sense of social responsibility of the enterprise.

References

Alonso-Garcia, J., Pablo-Marti, F., Núñez-Barriopedro, E., & Cuesta-Valiño, P. (2022). Digitalization in B2B marketing: omnichannel management from a PLS-SEM approach. *Journal of Business & Industrial Marketing*, (ahead-of-print).

Ant Group. (2022). *Ant Group 2021 sustainability report*. Hangzhou, China. https://gw.alipayobjects.com/os/bmw-prod/df8fcbe5-e60b-4ce7-9be4-55fb408b6f67.pdf

Cuesta-Valiño, P., Gutiérrez-Rodríguez, P., & Núnez-Barriopedro, E. (2022). The role of consumer happiness in brand loyalty: A model of the satisfaction and brand image in fashion. *Corporate Governance: The International Journal of Business in Society., 22*(3), 458–473. https://doi.org/10.1108/CG-03-2021-0099

Galán-Ladero, M. (2012). Variables que influyen en la actitud hacia el marketing con causa y determinantes de la satisfacción y la lealtad en la "compra solidaria". University of Extremadura. Doctoral Thesis. Cáceres: Servicio de Publicaciones de la UEx.

Normann, R., & Ramírez, R. (1993). From value chain to value constellation: Designing interactive strategy. *Harvard Business Review, 71*(4), 65.

Núñez-Barriopedro, E. (2019). Social Marketing in the General Directorate of Traffic's campaign called "Caminantedigital". In M. Galan-Ladero & H. Alves (Eds.), *Case studies on social marketing. Management for Professionals*. Springer.

Núñez-Barriopedro, E., Blanco González-Tejero, C., & Flores-Lario, L. (2021). Cause-related marketing (CRM) of Ausonia's campaigns. In *Cause-related marketing* (pp. 223–233). Springer.

Núñez-Barriopedro, E., & Llombart Tárrega, M. D. (2021). New trends in marketing aimed at the fourth sector in the fashion industry. In *Entrepreneurship in the fourth sector* (pp. 245–261). Springer.

Núñez-Barriopedro, E., Ravina-Ripoll, R., & Ahumada-Tello, E. (2020). Happiness perception in Spain, a SEM approach to evidence from the sociological research center. *Quality & Quantity, 54*(3), 761–779.

Pan, G. (2020). *The research on "Ant Forest" public welfare communication from the perspective of the interactive ritual chains*. Zhongnan University of Economics and Law. Master thesis. (In Chinese).

Penelas-Leguía, A., Galera-Casquet, C., Galán-Ladero, M., & Valero-Amaro, V. (2012). *Marketing solidario. El marketing en las organizaciones no lucrativas*. Pirámide.

Ravina-Ripoll, R., Nunez-Barriopedro, E., Almorza-Gomar, D., & Tobar-Pesantez, L. B. (2021). Happiness management: A culture to explore from brand orientation as a sign of responsible and sustainable production. *Frontiers in Psychology, 12*, 727845.

Zhang, Y., Xiao, S., & Zhou, G. (2020). User continuance of a green behavior mobile application in China: An empirical study of ant Forest. *Journal of Cleaner Production, 242*, 118497.

Part III

Sustainable Business

GreenBox: Ethics, Sustainable Management, and Social Commitment

Juana Edith Navarrete-Marneou
and Edgar Alfonso Sansores-Guerrero

Learning Objectives

1. Identify the consequences (environmental, social, and economic) of the pollution of the Mexican Caribbean coasts due to the use of plastics.
2. Understand, from the ethical perspective of corporate social responsibility (CSR), the social role of companies toward environmental problems.
3. Examine GreenBox's business model based on CSR.
4. Analyze the actions carried out by the GreenBox company to comply with its legal obligations, endorse its social commitment, and build a reputation.

1 Introduction

In the last decade, at a global level, the greatest environmental crisis has occurred due to several causes: the pollution of the oceans, the degradation of natural habitats, deforestation, the extinction of species, and the emission of toxic particles (Biodiversidad Arrecifal et al., 2020). This situation has caused an accelerated deterioration in the ecosystems which are considered essential for the survival of the human species and the conservation of the planet (Biodiversidad Arrecifal et al., 2020).

Among the natural resources that are most affected by pollution are the aquifers, which constitute the greatest wealth of biodiversity in the Yucatan Peninsula, specifically in the Mexican Caribbean. In this sense, the hyperconsumption of plastics (packaging, containers, utensils, and packaging, among others) has become one of the main causes of this problem, with approximately 13 million tons being dumped into the oceans annually (Reyes et al., 2019).

This large amount of garbage, when combined with other toxic agents and with temperature changes, has triggered a bacterial and algal increase that can be seen in

J. E. Navarrete-Marneou · E. A. Sansores-Guerrero (✉)
Universidad Autónoma del Estado de Quintana Roo, Chetumal, Quintana Roo, Mexico
e-mail: emarneou@uqroo.edu.mx; edsan@uqroo.edu.mx

E. C. Bianchi et al. (eds.), *Responsible Consumption and Sustainability*, Springer Business Cases, https://doi.org/10.1007/978-3-031-30742-3_8

the sargassum phenomenon in Quintana Roo (Reyes et al., 2019). Likewise, at the Peninsular level, industrial systems for the intensive production of pork and poultry have generated excessive volumes of waste (solid and chemical) that saturate the soil, aquifers, and air with their toxicity (Medina et al., 2021).

This situation has caused (a) losses to the tourism sector, (b) damage to the maritime ecosystem, (c) limitations to the human rights of access to quality water, (d) risks to public health, and (e) effects on social sectors, which are considered vulnerable (BBV, 2020).

Given these effects, organized civil society, higher education institutions, local producers, indigenous communities, and business organizations have expressed their concern in various forums, emphasizing the serious consequences to come in the upcoming years due to the high levels of pollution in the region and the eminent water crisis in Mexico (Comisión Nacional del Agua, 2019).

In this context, from the various sectors of society, the demand and commitment to address this problem through various mechanisms and campaigns have arisen. After an extensive discussion on the problem, these sectors have agreed on the need to generate a social policy for water management (restriction on the use of plastics, recovery of aquifers, promotion of circular consumption, restriction of polluting industries, and awareness, among others) that involves the participation of interested parties.

Given this, social organizations, government, and companies faced the great challenge of contributing, from their scope, resources, and institutional capacities to the attention of this social demand. Therefore, the solution to this environmental contingency would prevent an ecological disaster and a water crisis in the future (Biodiversidad Arrecifal et al., 2020).

2 Commitment and Actions of the Social Sectors

As part of these social agreements, in 2019, the government of the State of Quintana Roo, Mexico, assumed the commitment to create a regulatory framework to regulate solid waste management. As a consequence, the "Law for the Prevention, Comprehensive Management and Circular Economy of Waste" was enacted, which prohibits the use of single-use plastics and Styrofoam such as straws, plates, glasses, cups, trays, and cutlery, among others (Ley para la Prevención, Gestión Integral y Economía Circular de los Residuos del Estado de Quintana Roo, 2019).

Although the entry into force of said legal order intended to generate incentives for the eradication of the use of these materials, in the short term, it confused citizens since there was no strategy for its dissemination. There was a lack of advertising campaigns to raise awareness among the population about its benefits. Furthermore, the types of permitted materials were not defined, the seriousness of the pollution in the region had not been socialized, and the prices of biodegradable products were higher than plastic products (more than 40% [1]).

[1] Information provided by GreenBox, in 2020, based on quotes from its suppliers.

In these circumstances, companies together with nongovernmental organizations (NGOs) faced the commitment to design and implement innovative strategies aimed at promoting circular consumption among their customers, establishing awareness campaigns about caring for the environment, generating incentives for the acquisition of biodegradable products, and adopting ethical conduct among its consumers, implement social responsibility programs, and guarantee its investors the profitability of the business.

It was complex for companies to comply with these actions, especially for small ones, since they implied a paradigm shift in the way of conceiving the business-society relationship, as well as adopting a new relationship with their customers. Consequently, several companies that tried failed, obtaining a bad reputation in society due to a lack of ethics in handling information about the quality of their products or their high prices. On the other hand, businesses like GreenBox found an area of opportunity to position themselves in the market and build a solid reputation.

3 Genesis and Social Commitment of GreenBox

GreenBox, founded in 2018, arises to provide the restaurant and service industry with single-use plastic utensils, specializing in biodegradable ones. Since its inception, the company formulated as one of its value proposals "alternatives for circular consumption of single-use products," developing an environmentalist perspective.

In this sense, in 2020, the company designed and implemented a CSR program based on corporate ethics (Garriga & Méle, 2004). To do this, three objectives were set: (a) contribute to the conservation of the environment through responsible consumption, (b) endorse its social commitment to citizens, and (c) respect the laws and regulations of the State of Quintana Roo. From the objectives derived actions that were oriented toward the three dimensions of CSR: environmental, economic, and social (Table 1).

As part of its actions, GreenBox established a customer loyalty strategy. In addition, it increased its line of certified biodegradable products, implemented an eco-efficiency program in which the resources used within the company did not have a negative impact on the environment, redefined its organizational structure, and modified its infrastructure to guarantee access to all people, among others.

4 Transformation, Sustainability, and Responsibility

From its CSR program, GreenBox understood that its relationship with society would be based on shared ethical values, so the adoption of ethical practices in the management and marketing of its products represented its main challenge. This is coupled with their commitment to abide by the social and legal norms of their environment.

Consequently, the company was conceived as dependent on its context, which affected the definition of its purpose of assuming CSR as an ethical obligation above

Table 1 Corporate social responsibility actions

Dimension	Actions
Environment	Commercialization of certified biodegradable products Development of an awareness workshop for children in schools Awareness campaign on the effects of the use of nondegradable plastics Energy saving Program for the proper management of solid waste Promotion of a culture of circulating consumption Volunteer program for garbage collection in Chetumal Bay
Social commitment	Implementation of labor policies aimed at inclusion and respect for others Establishment of recruitment programs Establishment of fair wages Installation of a container for the collection of lids to help children with cancer
Concerned parties	Redefinition of the business model Establishment of a relationship system with the client Strict respect for legal regulations Training courses on CSR issues

Source: Own elaboration with information from GreenBox (2020)

any other consideration, including the economic one and participating in the attention of social and environmental problems (Brin & Nehme, 2019). Therefore, the redefinition of its management model toward sustainability was essential to generate, capture, and deliver added value and generate competitive advantages (Fig. 1).

In its model, GreenBox integrated three essential pillars: environment (environmental value and sustainable culture), customers (link), and sustainability (innovation, time). These were articulated through an awareness and training

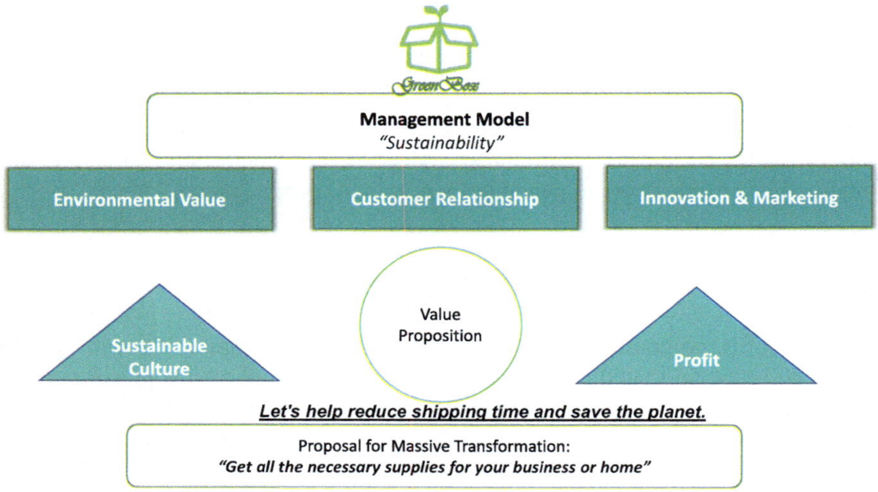

Fig. 1 GreenBox's management model. Source: GreenBox (2021)

Table 2 Relationship with interested parties

Stakeholders	Actions
Suppliers	It is guaranteed that they have certifications that are endorsed by law and international organizations that ensure that their products are of quality and that they meet the required biodegradability.
Government	A close relationship is maintained with the measures that are taken for the treatment of waste, as well as the effective application of the regulations that are taken concerning disposable products.
Customers	Our value proposition is based on three key elements: ensuring the quality of our products, minimizing delivery times, and working closely with our clients to produce engaging content for their social media platforms.
Society	A culture of circular consumption is promoted and there is participation in campaigns with a cause.

Source: Own elaboration from GreenBox's information (2020)

program in the use of biodegradable materials, the adoption of good practices in the management of the company's human resources, a customer loyalty program, and a CSR program.

Under this perspective, their bond with the client was seen as constant and sustainable, maintained by effective communication and a commitment to participate in those social causes aimed at meeting their demands, as well as distinguishing between what the client wants from what the client needs. For this reason, the redefinition of the product and attention to its externalities represented the starting point for each action.

The model endorsed the commitment to respect the environment and promote a culture of circular consumption where the customer was aware of the materials with which the products are manufactured and understood the need to conserve natural resources and the benefits of circular consumption. This involved a close relationship with stakeholders (Table 2).

5 Results

With the new management model and the CSR program implemented, the company fulfilled its three proposed objectives. Including digital social networks in its communication plan allowed it to generate valuable content and storytelling, based on shared values, to raise awareness of the need for circular consumption and the damages caused by the pollution of aquifers (Illustration 1).

A brand was built that constantly endorses its commitments and responsibilities established in a kind of social contract based on cultural and normative values shared with the citizenry, under a value proposal nuanced by ethics, commitment, and common values (Advantage, 2020).

In addition, organizational culture was developed based on the recognition of the other, inclusion, strict respect for human and labor rights, access to a decent life, and

Illustration 1 Awareness campaign. Source: GreenBox (2021)

constant training of personnel, which was essential because the company is growing, and human talent management is a pillar (Illustration 2).

Due to the above, GreenBox experienced unexpected growth, in times of pandemic (from 2019 to 2021), having a presence in two cities and two branches. Despite the crisis caused by the confinement measures implemented by the health authorities to reduce the spread of the SARS-CoV-2 virus, the company has positioned itself both in the market and in society due to its commitment, its respect for the laws, its ethical values, its participation in the care of causes, the quality of its products, and its social responsibility.

6 Conclusions

The implementation of actions aimed at establishing corporate social responsibility (CSR) programs represents an opportunity for companies to achieve their development and consolidate their link with the socioeconomic environment. The ethical perspective of business management assumes that business development is part of its main objectives since it seeks to obtain benefits that allow them to be sustainable, create a positive social image, implement institutional well-being, and comply with social norms. Consequently, CSR can become a means to achieve this purpose.

CSR is an alternative that provides benefits to the company, its environment, and interest groups, allows it to attract new customers, and positively influences the business image. In addition, it represents a strategy and a method of creating value for the business so that it is perceived by stakeholders as a company that contributes fairly and sustainably. CSR is based on the interrelationship between the company and society that shapes the economic-political reality, determines the degree of social integration, and promotes the exercise of ethical practices.

Illustration
2 Organizational culture.
Source: GreenBox (2021)

In this sense, social systems, as a concentrating entity, adapt to the conditions of their context for the fulfillment of collective objectives aimed at the integration and social cohesion of their members. Likewise, they systematically reproduce the cultural values of the society in which companies operate.

Sustainable management models have an important ally or a great threat in the social imaginary since the culture of environmental conservation is still very incipient and most consumers do not manage to make the benefits, maintaining a permanent behavior of respect for the environment, visible in the short term.

This culture, without a doubt, cannot be promoted exclusively through the products that people consume, so it is necessary to establish social responsibility actions that allow endorsing the social commitment of companies and integrating the attention of common problems, as well as adopting ethical behavior toward stakeholders.

GreenBox understood that the innovation of good business practices (ethics), the close link with society (commitment), and the flexibility to rethink and reinvent traditional business models based on sustainability are key elements for social transformation and the development of a culture of respect for the environment. However, there are still future challenges that GreenBox will have to face both at the commercial level and its social commitment.

7 Discussion Questions

1. What are the main consequences of pollution, due to the use of plastics, faced by the Mexican Caribbean?
2. What do you think of the role that companies play, through CSR programs, to address this environmental crisis?
3. Do you consider that the GreenBox business model is sustainable and responds to environmental problems? Why? Base your answer on the analysis of its components.
4. Do you consider that the actions carried out by the company contributed to improving its reputation? Why? What are the future challenges of GreenBox?

Appendix: Teaching Notes for the Case

Synopsis

The aquifers of the Mexican Caribbean in the Yucatan Peninsula, Mexico, from 2011 to 2021, have faced the greatest pollution in their history due to the hyperconsumption of plastics, the overexploitation of natural resources, and the intensive production systems. This situation has generated the demand of various sectors of society for attention to the problem itself. Therefore, the different actors in society developed a series of actions aimed at contributing to the reduction of pollution in the State of Quintana Roo, Mexico. The government enacted a law to regulate the consumption of plastic materials and some companies, together with social organizations, and implemented corporate social responsibility (CSR) programs as part of their commitment to society. From its scope, resources, and institutional capacities, each organization has contributed to generating a social agreement to avoid a water crisis. The objectives of this case study were the following:

(a) Identify the consequences (environmental, social, and economic) of the pollution of the coasts of the Mexican Caribbean due to the use of plastics.
(b) Understand, from the ethical perspective of corporate social responsibility (CSR), the social role of companies toward environmental problems.
(c) Examine GreenBox's business model based on CSR.
(d) Analyze the actions carried out by GreenBox company to comply with its legal obligations, endorse its social commitment, and build a reputation.

Potential Audience and Instructor's Material

This case has been developed for its use as teaching material in courses on corporate social responsibility, sustainable management, and social marketing.

The potential audience for this chapter is made up of the following:

(a) Undergraduate students of administration, marketing, or business programs
(b) Graduate students of business administration programs (MBA).

Learning Objectives

1. Identify the consequences (environmental, social, and economic) of the pollution of the Mexican Caribbean coasts due to the use of plastics.
2. Understand, from the ethical perspective of corporate social responsibility (CSR), the social role of companies toward environmental problems.
3. Examine GreenBox's business model based on CSR.
4. Analyze the actions carried out by the GreenBox company to comply with its legal obligations, endorse its social commitment, and build a reputation.

Time for Class Discussion

This case can be analyzed in a session of 60–90 min, depending on the number of students who participate with their contributions, as well as pre-class work.

Suggested Discussion

1. What are the main consequences of pollution, due to the use of plastics, faced by the Mexican Caribbean?
2. What do you think of the role that companies play, through CSR programs, to address this environmental crisis?
3. Do you consider that the GreenBox business model is sustainable and responds to environmental problems? Why? Base your answer on the analysis of its components.
4. Do you consider that the actions carried out by the company contributed to improving its reputation? Why? What are the future challenges of GreenBox?

Analysis

Answer to question 1. *What are the main consequences of pollution, due to the use of plastics, faced by the Mexican Caribbean?*

- This is a free-response question. (There is no right or wrong answer.) Each student expresses his/her opinion. The most relevant aspects of the question are found in the arguments that each student puts forward to understand the problem posed and assume a position in the face of this environmental crisis.

Answer to question 2. *What do you think of the role that companies play, through CSR programs, to address this environmental crisis?*

- Also, it is a free-response question. (There is no right or wrong answer.) Each student expresses his/her opinion based on the analysis of the case. So the key elements of the question are found in the arguments of each student to explain the role of companies to contribute to the solution of the problem through their CSR programs.

Answer to question 3. *Do you consider that the GreenBox business model is sustainable and responds to environmental problems? Why? Base your answer on the analysis of its components.*

- In the same way as the previous questions, this is a free-response question. (There is no right or wrong answer.) Each student expresses his/her opinion. The most relevant answer is the analysis of the components of the model based on the fundamentals of sustainability and CSR.

  Some ideas that could be developed in the question are the following:

- Characteristics of a sustainable business model
- Contribution to the attention of the detected problem
- Response to various interested parties

Answer to question 4. *Do you consider that the actions carried out by the company contributed to improving its reputation? Why? What are the future challenges of GreenBox?*

- Finally, this question is a free-response question (there is no right or wrong answer). Each student expresses his/her opinion. The main purpose is for students to relate the reputation of the company with the results obtained in the CSR program, as well as to identify the main future challenges of the company in the context of the problem.

References

Advantage, C. (2020). Corporate social responsibility. *CSR and Socially Responsible Investing Strategies in Transitioning and Emerging Economies, 65*.

BBV. (2020). *El riesgo del sargazo para la economía y turismo de Quintana Roo y México*. https://www.bbvaresearch.com/wp-content/uploads/2020/02/Riesgo_Sargazo_Big_Data.pdf

Biodiversidad Arrecifal, G. M. L., Oslet, R.-G. O., Lorenzo, Á.-F., Miguel, R., Ornela, G.-R., Esmeralda, P.-C., & Nuria, E.-S. (2020). Impacto de la contaminación por plástico en áreas naturales protegidas mexicanas. Greenpeace México. https://mexicosocial.org/wpcontent/uploads/2020/09/69659cfa-impacto-de-la-contaminacion-por-plastico-mexico.pdf

Brin, P., & Nehme, M. (2019). Corporate social responsibility: Analysis of theories and models. *EUREKA: Social and Humanities, 5*, 22–30.

Comisión Nacional del Agua. (2019). *Estadística del Agua en México*. http://sina.conagua.gob.mx/publicaciones/EAM_2019.pdf

Garriga, E., & Melé, D. (2004). Corporate social responsibility theories: Mapping the territory. *Journal of Business Ethics, 53*(1–2), 51–71.

GreenBox. (2020). *Programa de RSE*. Document consulted on December 20, 2021 at the company's facilities.

GreenBox. (2021). *Corporate Facebook profile*. https://www.facebook.com/GreenBoxChetumal1

Ley para la Prevención, Gestión Integral y Economía Circular de los Residuos del Estado de Quintana Roo (18 de Junio de 2019). XVI Legislatura. Periodico Oficial del Estado de Quintana Roo. https://www.congresoqroo.gob.mx/leyes/199/

Medina, L., Fernández, J., & Montiel, J. (2021). Contaminación del Acuífero Maya. https://dplf.org/sites/default/files/informe_contaminacion_acuifero_maya.pdf

Reyes, H., Álvarez, L., Pérez, H., & Santillo, D. (2019). *Estudio sobre el impacto de la contaminación por microplásticos en peces de México*. Greenpeace. https://www.researchgate.net/publication/339693389_Estudio_sobre_el_impacto_de_la_contaminacion_por_microplasticos_en_peces_de_Mexico

Corporate Social Responsibility (CSR) and Sustainable Production in the Wine Industry: "Familia Torres"

Estela Núñez-Barriopedro, Blanca García-Henche, and Mariam Tkhinvaleli

Learning Objectives

1. Deepen the role of corporate social responsibility and learn how it has been applied in the wine industry.
2. Analyze what kind of projects of corporate social responsibility and sustainability they have had and what kind of results they have achieved.
3. Analyze the positioning and differentiation of Familia Torres in the wine industry through corporate social responsibility strategies and sustainable production.
4. Find out in general what kind of objectives the company has in relation to sustainable production and what actions it offers against different types of problems around the world.
5. Analyze the usage of their communication campaigns, especially in digital channels, and see how engaged they are with the people interested in their actions.

1 Introduction

Familia Torres owns over 1300 hectares of vineyards in Spain, which are cultivated with different grape varieties. In its native Catalonia, the Ancestral Varieties Project has recovered Forcada, Querol, Gonfaus, Pirene, and Moneu, five previously unknown indigenous grapes with exceptional potential for producing high-quality wines, in addition to the well-known Carignan, Grenache, Monastrell, Parellada, Sumoll, and Xarello varieties. Torres grows both local (Albario in Ras Baixas, Tinto Fino in Ribera del Duero, Tempranillo in Rioja) and foreign wine varieties (Cabernet Sauvignon, Cabernet Franc, Syrah, and Riesling) in the rest of Spain (Familia Torres, 2018). Thus, "Familia Torres" markets four main product lines, as shown in Table 1.

E. Núñez-Barriopedro (✉) · B. García-Henche · M. Tkhinvaleli
Universidad de Alcalá, Alcalá de Henares, Madrid, Spain
e-mail: estela.nunezb@uah.es; blanca.garcia@uah.es; mariam.tkhinvaleli.1@iliauni.edu.ge

E. C. Bianchi et al. (eds.), *Responsible Consumption and Sustainability*, Springer Business Cases, https://doi.org/10.1007/978-3-031-30742-3_9

Table 1 Main product lines in "Familia Torres"

Brand	Differentiation	Location	Characteristics
Pago del Cielo	DO Ribera del Duero	Fompedraza	Ideal climatology for the production of red wines with great aromatic intensity
	DO Rueda	Villafranca del Duero	Exceptional location for the production of verdejo grapes
La Carbonera	Classified as a singular vineyard by the DO La Rioja	La Rioja	Focuses on origin and typicity to produce unique wines that interpret the landscape of Labastida and the Rioja Alavesa
Pazo Torre Penelas	DO Rías Baixas	Portas	Focused on the production of unique wines in innovative ovoid granite vats from the walled vineyard
Vardon Kennett	High-quality sparkling wines	Mediona, Alt Penedès	They are made with pinot noir, chardonnay and Xarello from their own vineyards located at an altitude of 500 m above sea level. After the harvest, alcoholic fermentation is carried out with indigenous yeasts.

Source: Own elaboration based on the information published in the "Familia Torres" website (2022)

Supply chain management is fundamental not only to ensure product quality but also to extend "Familia Torres" values and beliefs to all those involved in the commercialization process. For this reason, it has procedures and policies that help it to identify and manage risks. In this way, the supplier policy is based on a model of ethics and social responsibility, where the general framework is defined to ensure the minimum requirements that suppliers must meet in order to form part of the supply chain. At the time of approval, this policy applies to all of them (Rovira, 2021).

2 Case Development

Although to some people "Familia Torres" may seem like a normal wine-making company, this is not the case. It is a company that is interested in making the work experience as useful and possible not only in the field of wine but also in the field of world problems, where lots of attention is focused from the side of the Torres winery.

Familia Torres aspires to be a source of prosperity, employment, and opportunity, so it is passionately committed to the communities and environments in which it operates. As a result, one of its best practices is to encourage the hiring of local suppliers while still preserving successful business partnerships that benefit both the company and the community.

Familia Torres is aware that some parts of the world are more likely to violate human rights and have a negative influence on the environment. Therefore, it has implemented more severe systems to recognize all hazards that, due to their severity, could have substantial ramifications on the company's operations and reputation. So whenever a supplier manufactures in an at-risk country, an external organization conducts an audit (Asia Inspection). Operations, corporate social responsibility, the environment, cleanliness, and health and safety are all parts of the audit. Inspections must be conducted at the factory, at the beginning, middle, and end of production, in all audits. Asia Inspection gives the final reports to the supplier as well as Familia Torres' Purchasing Department (Familia Torres, 2018). Therefore, all the campaigns and collaborations in which the company has been involved have left their mark in all areas of CSR.

"Familia Torres" can be considered as a marketer with over a century of experience in the traditional production of wine, which is the firm's primary source of value. The company also produces value by keeping the classic and aged features that set it apart from the competition (Cuesta-Valiño et al., 2020).

"Familia Torres" fosters the production of shared value in society through social acts and participation in local and/or industry organizations as part of its commitment to social responsibility. The goal of these collaborations is to play a more active role in the sector's growth by promoting the places in which it works and enhancing the environment in which it functions. It also attempts to promote discussion on the current difficulties that businesses face (Ravina-Ripoll et al., 2019).

The "Familia Torres" case contributes to meeting social needs from two central focuses that will be developed below. Firstly, it contributes to sustainability by collaborating with climate action through the actions of the "Torres & Earth" program. Secondly, it develops a variety of CSR projects, with the "Vyu la Vynia" project in the fight against disabilities being one of the most important projects in which it participates through its foundation.

2.1 Description of the Collaboration of "Familia Torres" with Climate Action Through the Project of "Torres and Earth"

In Spain, the International Wineries for Climate Action (IWCA), created in 2019 by Familia Torres from Spain and Jackson Family wines from California (USA), is worth mentioning. This organization's objective is to take collective action to decarbonize the global wine business by meeting ambitious and measurable CO_2 emission reduction targets (IWCA, 2019). And in 2021, they created, with four other partners, the first Regenerative Viticulture Association[1] in Spain.

The objective to be achieved is that all associated wineries have a long-term strategy to reduce carbon emissions by 80% by 2045 and with a shorter-term goal of 50% by 2030. To achieve this, it involves investing in renewable energy, working in the three areas of emissions reduction, and having a CO_2 emission reduction track record of at least 25%, confirmed by a certifying authority.

---

[1] The association was founded by the following five family wineries: Familia Torres, Clos Mogador, Can Feixes, Jean Leon, and the agricultural consultancy AgroAssessor.

The Torres & Earth sustainability initiative was established with the goal of lowering CO_2 emissions by 30% per bottle between 2008 and 2020 from the vineyard to final logistics. The business is stepping up its efforts to lessen the effects of the climate emergency and adapt to them. Due to this, the business has set a new goal for itself: by 2030, it wants to cut its CO_2 emissions per bottle by 55% compared to 2008 levels.

Familia Torres has always tried to achieve success without harming the Earth. Faced with a global trend of rising temperatures, the company decided to step up its efforts to care for the land and conserve the environment, designating the battle against climate change as one of the key axes of its environmental strategy (Familia Torres, 2022). Beginning in 2020, the vineyard stepped up its efforts to adapt to and mitigate the effects of change, with new goals in mind: cut direct and indirect CO_2 emissions by 55% in 2030 and 80% in 2045, compared to 2008 (Familia Torres, 2022).

Emissions have been reduced by the following main factors: the use of renewable energies, energy saving techniques, environmentally friendly transport, and the emergence of lighter bottles. Thus, the choice of suppliers with a low carbon footprint has significantly reduced the winery's carbon footprint. Specifically, suppliers of raw materials, packaging, and distribution account for approximately 90% of the winery's carbon footprint (Winetitles, 2022).

This company invests resources in research to find ways to adapt to and mitigate climate change. One of the areas in which it works is the development, in collaboration with universities and other companies, of technologies for capturing and reusing CO_2, known as CCR (carbon, capture, and reuse) technologies. It also participates in and leads national and international projects to study the influence of climate change on vineyards and wine and to develop new adaptation and mitigation mechanisms.

Four new photovoltaic projects will be installed in 2022 and 2023, allowing the winery to become 50% self-sufficient in its energy supply, in addition to solar panels and a biomass boiler that have been in service for a number of years.

To reach its goal of reducing direct and indirect CO_2 emissions by at least 60% by 2030 when compared to 2008 and becoming a carbon-neutral winery before 2040, the winery will use self-generated renewable energy (Yates, 2022).

2.2 "Familia Torres" Against Disabilities

The CSR initiatives embodied in each of the projects in which the "Familia Torres" participates have served as a model for other companies in the region. Torres has been chosen on numerous occasions as the "most admired wine brand in the world" by specialists in the sector, demonstrating a track record of successful business management (Toribio, 2020).

Such success of the Familia Torres company is also due to its attitude of unwavering support to those who participate in the program. This assistance is feasible thanks to the coordination with several foundations that act as a link between the company and the social needs. The CSR projects carried out by the company in Europe are summarized in Table A1 in Appendix 1.

In this sense, it is worth mentioning that several of the workers of the Mas Albornà Foundation have become employees of Familia Torres. In this way, it supports the creation of jobs for people with disabilities and their integration into the labor market. There are people working at the Penedès facilities who, because of their intellectual disability, have more difficulties in finding a normal job. The number of workers fluctuates monthly depending on operational needs and the seasonality of the products; they work in product handling processes (logistics, labeling, assembly of promotions, etc.) and, to a lesser extent, in kitchens, in catering jobs, and in the development of an educational workshop for schools.

In addition, Familia Torres promotes the Viu la Vinya project, a center for knowledge and dissemination of the world of the vineyard, which is located in its own vineyard and is run by people with disabilities, in collaboration with the Mas Albornà Foundation. In this way, visitors have a pleasant and practical experience exploring the vineyard environment (harvesting grapes, making cuttings, experimenting with renewable energies, learning environmentally responsible attitudes, etc.), while normalizing diversity and appreciating its added value (Ravina-Ripoll et al., 2021).

According to Familia Torres, these projects have enormous social significance for people with disabilities because they have contributed to eliminating stigma and social labels, making them feel respected for their abilities and functions.

Ethical motivations have become one of the main drivers of purchasing decisions for products and services (Andreasen, 2003; Cuesta-Valiño et al., 2019).

3 The Main Aims of "Familia Torres"

From the beginning in 1986, the Familia Torres Foundation has supported over 400 international cooperation initiatives with the following key goals:

- Child protection through the building of educational institutions and foster homes for orphaned and vulnerable children.
- Women's empowerment and the battle against gender violence.
- Improved health among special needs groups. People should be exposed to culture.
- Humanitarian relief and assistance in disaster-stricken areas.

In 2018, the Familia Torres Foundation received a total amount of 436,968.68 euros in donations and established 388,947.98 euros in social action collaborations. Child protection (127,721.95 euros), help to groups with special needs (142,732.22 euros), and assistance to women and families were among the areas to which it paid special attention (80,248.34 euros). So "Familia Torres" has donated the funds to the specific organizations. The results of a research on donation destinations are presented in Table 2.

Table 2 The organizations where "Familia Torres" donated funds

Asociación Empresarial de Bodegas D.O. Ribera del Duero	Associació Vinícola Catalana	Consejo Empresarial Alianza por Iberoamérica	Consell Català Prod. Agrària Ecològica
Spanish Wine Federation	Gleve	Greenpeace Spain	Grupo de Empresas Vinícolas de Rioja
Grupo Español para el Crecimiento Verde	Instituto de la Empresa Familiar	Interprofesional del Vino de España	Primum Familiae Vini, S.L.
Qalides Terrers del Penedès	Transprime	Union de Licoristas de Cataluña	World Compliance Association

Source: Own elaboration

4 Results of the Campaigns and Collaborations

The main results of the most significant campaigns are as follows:[2]

The collaboration of "Familia Torres" *with climate action.* Wineries must use 15% self-generated renewable energy, have reduced their carbon footprint by 15%, and have a 35% reduction goal by 2030.

As one of the fundamental pillars of its environmental policy, they advocate a series of measures to adapt to and mitigate climate change. From the vineyard through final logistics, they have reduced CO_2 emissions per bottle from 2008 by 30% to 2020. Since 2020, the vineyard has increased its efforts to adapt to and mitigate the effects of change, with new goals in mind, including a 55% reduction in direct and indirect CO_2 emissions in 2030 and an 80% reduction in 2045 (Familia Torres, 2022).

This company has a YouTube channel where it broadcasts videos of its campaigns. On the front page, the video entitled "We celebrate life, we care for the earth and carry on our legacy" can be seen with a total of 3423 views.

"Torres and Earth" program exists in the "Familia Torres," so they play their part in taking care of the planet Earth, which is very difficult, when the topic is about having a factory. "Familia Torres" managed to reduce the air pollutant emissions, toxic waste disposal, and water contamination, so the Earth could "breathe" a lot better than before.

Among the promotional campaigns in which this initiative is disseminated, we can find on the YouTube channel under the slogan "The more we take care of the land, the better wine we get"[3] with a total of 1054 views.

"Familia Torres" is against disability through the "Vyu la Vynia" *project.* This initiative helps these people with disabilities to increase their happiness and

[2]Exploratory research was made. The revision of secondary sources from the literature and online resources of the entity was fully analyzed.

[3]https://youtu.be/2RyneEmr_ow

Table 3 Social media activity

Social media	Instagram	Facebook	LinkedIn	Twitter
Subscribers	12.7 K	970,262	20.395	25.8 K
Average quantity of comments	6–7	200–300	5–6	5–6
Average quantity of likes	150–160	10 K	30–35	26.8 K
Quantity of posts	792	52	84	4978

Source: Own elaboration

satisfaction in life, as well as with their co-workers, by implementing an inclusive work life in which disabilities are normalized (Núñez-Barriopedro et al., 2020).

Among the promotional campaigns in which this initiative is disseminated, we can find under the slogan "For the well-being of people, promoting education and social integration"[4] with 46 views.

"Familia Torres" is also quite active when it comes to social media. They are passionately dedicated to sharing their news on the social media platforms, so their users always find out the news about them first. In the yearly, monthly meetings or presentations that they have, they make sure to document it and share it with their followers (see Table 3).

Based on the value chain, the after-sales service is where more value is generated, because this information is obtained from customers, allowing changes to be made within some processes and providing customers with a product that they can perceive its value, as it is sought that the product be what you want or expect. This after-sales service is supported by the company's online platform, which provides benefits to its customers through advertisements, wine information, and an online support network with company experts to ensure that consumers fully understand the brand or product to be consumed. This service will build consumer loyalty.

The market research should not be centered on gathering information about clients that a company already has and knows, because most of them are familiar with the classic wines produced, making it difficult for them to accept another wine just because it follows market trends. As a result, data should be gathered from possible new product users, beginning with customers who drink wine for enjoyment, health, and not to become drunk, such as pregnant women, athletes, and anyone searching for a drink, participating in social occasions while maintaining a healthy lifestyle (Gilinsky et al., 2016).

5 Conclusions

"Familia Torres" is a successful winemaking company, which is developing many processes with the aim of trying to move toward a sustainable future with the help of new technologies.

[4]https://youtu.be/j2n9lmi2E0w

There are not many companies in the world, which try to make the business successful, while at the same time trying to worry about the environment and the current situation on the planet Earth.

Familia Torres is dedicated to generating shared benefit, and its tax structure reflects its values of transparency and ethics. None of its businesses are located in tax havens as a result of this philosophy. The tax strategy and policy for controlling and managing tax risks are determined by management in collaboration with the board of directors. It also ensures the accuracy of tax data sent to stakeholders, as well as adhering to the legislative requirements of the nations in which it operates and acting in the best interests of society.

In fact, companies market their products, in this case wine, in a globalized environment, and companies are dynamic entities linked to the environment. Therefore, there is a company-society relationship that has to be aligned with the objectives, interests, and ethical issues. Therefore, there are stakeholders that provide a competitive advantage to the company,[5] who rely on the adoption of social responsibility by the company through socially responsible initiatives, with which companies manage to improve their image and achieve an emotional differentiation, based on values, establishing longer-term relationships with their customers.

6 Discussion Questions

1. Do you consider that the projects carried out by the "Familia Torres" really contribute to sustainability and the benefit of society or is it a strategy to gain brand value?
2. What is the role of online communication strategies in the "Familia Torres" campaigns?
3. Through CSR strategies of the "Familia Torres," do you think it has managed to differentiate itself in the wine industry from other designations of origin (DO) and how would it position itself?

[5] As stated by authors such as Galán-Ladero et al. (2004) and Galán-Ladero (2012).

Appendices

Appendix 1

Table A1 Corporate social responsibility projects carried out by Familia Torres in Europe

Location	Year	Project title	Features
Spain	Annual	La Marató TV3	The Fundació La Marató TV3 was created in 1996 by the Catalan Corporation of Audiovisual Media with the mission to encourage and promote biomedical research of excellence, as well as to raise social awareness of the diseases addressed in the television program, La Marató de TV3, through citizen participation campaigns and dissemination and educational events. Through the "Familia Torres" Foundation, we participate annually in a day of solidarity experiences paired with the world of wine.
Spain	2021	Daydreaming Foundation —Educating smiles	This project was created to improve the academic and relational situation of minors living in 24 educational action residential centers (CRAEs) in Catalonia with the aim of guaranteeing an adequate school level for their development and future social integration, acquiring the necessary knowledge, study habits, and social skills.
Spain	2020	Actua Vilafranca—Ecological vegetable garden for social inclusion	The children of the CRAE Toni Inglés, located in the San Julián neighborhood of Vilafranca, take charge of a vegetable garden in a pilot project of agriculture and teamwork.
Spain	2020	Mas Albornà Foundation—Just like at home	Adequacy of the home-residence space to improve the quality of life of the residents. The Mas Albornà residence located in Las Cabañas is a seventeenth-century farmhouse converted into a residential space for people with intellectual disabilities.
Spain	2020	Fundación Grupo Àuria—Children of the Future	Children of the Future is a collaborative project that creates a natural meeting place for people with disabilities and children from the social environment of the municipality and region.
Spain	2020	Bobath Foundation—Bobath Shop	The Bobath Shop project of the Bobath Foundation in Madrid was born in 2014 with the aim of developing to the maximum the sensorimotor, cognitive,

(continued)

Table A1 (continued)

Location	Year	Project title	Features
			and social skills of the users of the Day Center, from the approach of occupational therapy. The Bobath Foundation is the pioneer welfare entity in Spain in the comprehensive treatment of children and adults affected by paralysis or brain damage.
Spain	2020	Associació Punt de Referència—Proyect Acoge	Punt de Referència offers a temporary shelter to immigrant teenagers in a family unit, which lasts nine months. The young person is offered to live in a family environment with people culturally rooted in Catalonia.
Spain	2020	Actua Vilafranca—Project families	Psychological and psychosocial care project for families, children, and adolescents of CRAE Antoni Anglès, managed by the Actua Vilafranca. The contribution has allowed to incorporate during this 2019 the figure of a psychologist to work in coordination with the other educators for 12 h a week.
Spain	2020	Barcelona Actua—Project "Dones del Món"	Project aimed at 18 mothers in a situation of social vulnerability and economic precariousness, derived from social services.
Spain	2020	ARED Foundation—Reintegration of women	The foundation works to achieve the social and labor integration of people in a situation of social exclusion, mainly women from prisons and social services.
Spain	2018	Font de la Canya—Arqueovitis Cooperativa	Archaeological days: A journey through time that allowed about 60 collaborators and their families to learn about the cradle of viticulture in the Penedès.
Spain	2018	Actua Vilafranca—Project for the reception of unaccompanied minors	Project of socio-educational intervention with children and young people without family references (MENAS).
Spain	2018	Barcelona Actúa—Becas D-Bac	This project aims to help their employability through training and professional support. The women candidates are users of "Dones del Món."
Spain	2018	Barcelona Actúa—Becas U-BAC	The objective is to promote the social elevator through education, in young people from families at risk of social exclusion, who belong to high-complexity institutes of Barcelona.

(continued)

Table A1 (continued)

Location	Year	Project title	Features
Belgium	2012	Le Village n° 1	The foundation made a financial contribution to a project for the construction of a house inside the educational center for children with physical and psychological problems that the Village No. 1 Foundation has in operation. It consists of a project to help and reintegrate these children into society.
Spain	2004	Aldeas Infantiles—Construction of shelters	The project is carried out through a financial contribution to the organization Aldeas Infantiles-SOS and is materialized in the construction of shelters for 50 children from dysfunctional families, victims of abuse, or orphans.

Source: Own elaboration from the "Familia Torres" foundation website (2022)

Appendix 2: Teaching Notes for the Case

Synopsis

"Familia Torres" is a Spanish wine producer and marketer with over a century of experience in the traditional production of wine. Exports have been one of the company's key cornerstones since its inception, and "Familia Torres" wines can today be found in over 150 countries.

The whole organization is oriented on creating biological products. "Familia Torres," as an environmentally aware corporation, is particularly worried about how climate change would affect the wine industry. As a result, one of the company's programs is "the Torres & Earth Program," created to implement actions that assist in limiting the effects of climate change.

One of the main collaborations of the company is the IWCA (International Wineries for Climate Action), which is a collaborative working group that addresses climate change through creative carbon reduction measures. "Familia Torres" meets a set of criteria, which applies to promoting sustainability through four pillars: greenhouse gas (GHG) reduction, water management, waste reduction, and energy efficiency and renewable energy promotion.

Also "Familia Torres" manages to deal with the other widely spread problems. For example, since 1988, Familia Torres has worked with the Mas Albornà Foundation, which promotes the employability of people with intellectual disabilities.

Potential Audience and Instructor's Material

This case study can be used as study material for marketing and marketing management courses. Because this case study takes into account social development, business models, etc., it can also be used in several other courses, such as economics, sociology, regional development, marketing, etc.

Learning Objectives

1. Deepen the role of social marketing and learn how it has been applied in the wine industry.
2. Analyze what kind of projects of social marketing they have had and what kind of results they had.
3. Analyze the positioning and differentiation of "Familia Torres" in the wine industry through social marketing strategies and sustainable production.
4. Find out in general what kind of objectives the company has in relation to sustainable production and what actions it offers against different types of problems around the world.
5. Analyze the usage of their social media and see how engaged they are with the people interested in their actions.

Time Frame for Class Discussion

This case can be taught in one session of approximately 45 min.

Suggested Discussion Questions

1. Do you consider that the projects carried out by the "Familia Torres" really contribute to sustainability and the benefit of society or is it a strategy to gain brand value?
2. What is the role of online communication strategies in the "Familia Torres" campaigns?
3. Through the social marketing strategies of the "Familia Torres," do you think it has managed to differentiate itself in the wine industry from other designations of origin (DO) and how would it position itself?

Analysis

Answer to question 1: *Do you consider that the projects carried out by the "Familia Torres" really contribute to sustainability and the benefit of society or is it a strategy to gain brand value?*

One of the most successful collaborations that "Familia Torres" had is "Torres and Earth," because the planet Earth is facing lots of pollution and growing problems since the end of the twentieth century. "Familia Torres" is a part of a campaign, which is trying to stop all these things, which will make our planet breathe freely. That is something that not many companies and factories are worried about. It is a very good project to make wine and make it in a healthy way, which also increases the trust level in the consumers' mind. Whenever a cause is relevant to the consumer, he/she will feel more interested and involved with that action (Berger et al., 2006), so the feedback will be more favorable (Gupta & Pirsch, 2006) and, consequently, purchase intention will increase. Another relevant project is "Familia Torres" against disabilities "Vyu la Vynia" project, because this initiative helps these people with disabilities to increase their happiness and satisfaction with life, as well as their co-workers by implementing an inclusive working life (Núñez-Barriopedro et al., 2020).

This is a win-to-win formula. On the one hand, the world that needs help in terms of sustainability and pollution reduction wins, and the company that has created the campaign, in this case "Familia Torres," wins. Finally, the consumer also wins by contributing to a good cause without any extra outlay, simply by buying the wine. For all these reasons, it is a formula that allows the realization of projects that otherwise would not achieve the same visibility and consumer appreciation, which indirectly also benefits the brand (Núñez-Barriopedro et al., 2021).

Answer to Question 2: *What is the role of online communication strategies in the "Familia Torres" campaigns?*

"Familia Torres" will be able to contact the consumers and tell them of the situation as well as provide alternate solutions. Familia Torres' approach to communication and transparency, two critical aspects in customer happiness, is reaffirmed by this 360° customer service management.

"Familia Torres" is also quite active when it comes to the social media. They are passionately dedicated to sharing their news on the social media platforms, so their users always find out the news about them first. The yearly, monthly meetings or presentations that they have, they make sure to document it and share it with their followers.

Therefore, practitioners should bear in mind that effective communication must vary the style and content of the message when it comes to communicating healthy food, specifically sustainable wine. Credibility, informational and entertainment value are key attributes for assessing the value of advertising about these wines on social media. It is important to offer content that engages through credibility. In addition, credibility is its own nature because online media are the most credible means of obtaining information for consumers (even social media). So, "Familia Torres" should try to show videos or images about the brand through communicating its social marketing and sustainable and ecological production projects, or use news written and hired by journalists and media professionals, but even traditional banner advertising could attract this kind of consumers if we focus on credibility and information (Cuesta-Valiño et al., 2020).

Answer to question 3: *Through CSR strategies of the "Familia Torres," do you think it has managed to differentiate itself in the wine industry from other designations of origin (DO) and how would it position itself?*

"Familia Torres" has been operating in the winery segment for a really long time, so their job and their product (judging from the research carried out) is one of the most successful winery products in Spain. Torres has been voted "World's Most Admired Wine Brand" by industry specialists on numerous occasions, proving that it is indeed a success story.

Through each social marketing project, it has achieved recognition in the wine market as a responsible and sustainable industry with a differentiation that is committed to organic wine, care for the environment, and social inclusion in the labor market.

The wine market presents a complex, changing, and highly competitive panorama. Therefore, this case study analyzes the initiatives of "Familia Torres" that implement social marketing strategies creating an added value from the differentiation in the denominations of origin and the happiness of its workers for a better positioning in the wine market both in the national and international markets (Barriopedro et al., 2019).

References

Andreasen, A. R. (2003). The life trajectory of social marketing some implications. *Marketing Theory, 3*(3), 293–303.

Barriopedro, E. N., García, I. P., & Valiño, P. C. (2019). Estrategias de marketing desde la diferenciación en las denominaciones de origen, la creatividad y la felicidad para un mejor posicionamiento en el mercado del vino. In *Educación y felicidad en las ciencias sociales y humanidades: un enfoque holístico para el desarrollo de la creatividad en la era digital* (pp. 83–102) Tirant lo Blanch.

Berger, I. E., Cunningham, P. H., & Drumwright, M. E. (2006). Identity, identification, and relationship through social alliances. *Journal of the Academy of Marketing Science, 34*(2), 128–137.

Cuesta-Valiño, P., Rodríguez, P. G., & Núñez-Barriopedro, E. (2019). The impact of corporate social responsibility on customer loyalty in hypermarkets: A new socially responsible strategy. *Corporate Social Responsibility and Environmental Management, 26*(4), 761–769.

Cuesta-Valiño, P., Rodríguez, P. G., & Núñez-Barriopedro, E. (2020). Perception of advertisements for healthy food on social media: Effect of attitude on consumers' response. *International Journal of Environmental Research and Public Health, 17*(18), 6463.

Familia Torres. (2018). *Corporate social business report.* Home | Familia Torres Las viñas de Torres vuelven al origen – EL PAÍS (singulardigital.mx)

Galán-Ladero, M. (2012). *Variables que influyen en la actitud hacia el marketing con causa y determinantes de la satisfacción y la lealtad en la "compra solidaria"*. Doctoral Thesis, University of Extremadura. Cáceres: Servicio de Publicaciones de la UEx.

Galán-Ladero, M., Galera-Casquet, C., & Valero-Amaro, V. (2004). La convergencia del Management y del Marketing en el ámbito de las causas sociales. Una referencia al caso español. *Investigación y Marketing, 85*, 52–62.

Gilinsky, A., Jr., Newton, S. K., & Vega, R. F. (2016). Sustainability in the global wine industry: Concepts and cases. *Agriculture and Agricultural Science Procedia, 8*, 37–49.

Gupta, S., & Pirsch, J. (2006). The company-cause-customer fit decision in cause-related-marketing. *The Journal of Consumer Marketing, 23*(6), 314–326.

IWCA. (2019). *International wineries for climate action.* https://www.iwcawine.org/

Núñez-Barriopedro, E., Ravina-Ripoll, R., & Ahumada-Tello, E. (2020). Happiness perception in Spain, a SEM approach to evidence from the sociological research center. *Quality & Quantity, 54*(3), 761–779.

Núñez-Barriopedro, E., Blanco González-Tejero, C., & Flores-Lario, L. (2021). Cause-related marketing (CRM) of Ausonia's campaigns. In *Cause-related marketing* (pp. 223–233). Springer.

Ravina-Ripoll, R., Núñez-Barriopedro, E., Evans, R. D., & Ahumada-Tello, E. (2019). Employee happiness in the industry 4.0 era: Insights from the Spanish industrial sector. In *2019 IEEE technology & engineering management conference (TEMSCON)* (pp. 1–5). IEEE.

Ravina-Ripoll, R., Nunez-Barriopedro, E., Almorza-Gomar, D., & Tobar-Pesantez, L. B. (2021). Happiness management: A culture to explore from brand orientation as a sign of responsible and sustainable production. *Frontiers in Psychology, 12*.

Rovira, M. (2021) *Las viñas de Torres vuelven al origen*. Las viñas de Torres vuelven al origen – EL PAÍS (singulardigital.mx)

Toribio, C. (2020). Cómo construir conjuntamente VALOR en Gran Consumo. In *27° Congreso AECOC de estrategia comercial y marketing*. AECOC | La asociación de fabricantes y distribuidores|AECOC | La asociación de fabricantes y distribuidores

Torres. (2022). https://www.torres.es/sostenibilidad-ambiental

Winetitles. (2022). *Winetitle media*. https://winetitles.com.au/familia-torres-consolidates-its-winemakingproject-in-galicia-with-the-purchase-of-a-new-winery/

Yates, E. (2022). https://drinksint.com/news/fullstory.php/aid/10135/Familia_Torres_set_to_self-supply_50_25_of_energy_at_its_Pacs_del_Pened_E8s_winery_by_2023.html

Providing Sustainable Solutions in Chile: Neptuno Pumps—Innovation That Flows

Luis Manuel Cerda-Suarez, Iria Paz-Gil, and Dolores Lucia Sutil-Martin

Learning Objectives

1. To identify responsible consumption and production trends in a sustainable business model in Chile.
2. To illustrate the importance of the efficient and client-customized use of natural resources in one of the most arid places on the planet, the Atacama Desert.
3. To explore what marketers should focus on when designing sustainable business models that promote the circular economy (CE) in several regions and places.
4. To describe a relevant initiative in the pump industry that delivers innovative, sustainable, and efficient solutions through recycling and reuse.

1 Introduction

The 2030 Agenda, approved by the United Nations General Assembly on September 25, 2015, is an integrative plan of action conceived as a tool for tackling the greatest global challenges we face and is a requirement for sustainable development (Lubowiecki-Vikuk et al., 2021). The Sustainable Development Goals that it sets out are integrative and address the three main dimensions of sustainable development: the social, economic, and environmental (Bhattacharyya et al., 2021). With this in mind, driving actions to implement a new circular business model by integrating industry 4.0 technologies and thereby contributing to the achievement of the SDGs and SDG 12 in particular—responsible consumption and production—is of particular relevance worldwide (Akehurst et al., 2012).

L. M. Cerda-Suarez (✉)
International University of La Rioja, Logroño, La Rioja, Spain
e-mail: luis.cerda@unir.net

I. Paz-Gil · D. L. Sutil-Martin
University Rey Juan Carlos, Madrid, Spain
e-mail: iria.paz@urjc.es; doloreslucia.sutil@urjc.es

E. C. Bianchi et al. (eds.), *Responsible Consumption and Sustainability*, Springer Business Cases, https://doi.org/10.1007/978-3-031-30742-3_10

Although the concept of circular economy (CE) is most often discussed, both academically and professionally,[1] in terms of sustainability, CE models also offer alternatives to traditional linear business models. In CE models, final production is also the source of value creation. This delivers products and items with longer life cycles in accordance with three main objectives (Altinbasak-Farina & Burnaz, 2019): (1) to preserve natural capital by balancing resource flows; (2) to encourage the efficient use of resources by better distributing products, components, and materials that can be used in natural cycles; and (3) to increase the effectiveness of the system by reducing negative externalities.

Regarding its relevance to the SDGs, CE has the combined aim of fighting climate change, creating new opportunities for business, and generating high-quality new jobs (Cuevas, 2016). When designing a new business model that could incorporate technological innovations and contribute to the achievement of the SDGs, Neptuno Pumps worked with its strategic partners and clients in Chile to develop a sustainable and innovative circular model which produces pumps that are up to 1000% more reliable and up to 30% more efficient, uses material recycled from scrap equipment for 60% of company goods, eliminates waste by up to 75%, and reduces the company's carbon footprint by 70% (Kowszyk & Maher, 2018). These figures illustrate the benefits that the company has generated at international and national level since adopting a model geared toward recycling thousands of cubic meters of water that was previously wasted across Latin America's industrial sectors, recovering and reusing a large amount of material that was previously thrown away, and reducing carbon emissions from operational processes and partner and client activities.

In pioneering the circular economy, Neptuno Pumps is aligned with the new business reality that has its roots in innovation[2] and sustainability.[3] Neptuno Pumps has vertically integrated manufacturing processes that demonstrate just how important it is to remanufacture worldwide as they enable the company to offer a range of goods that are up to 30% cheaper and 25% more energy efficient and that have the same guarantees as new industrial equipment. Neptuno Pumps forms new alliances with clients and partners and works with them to reach its goal of making 90% of its products from recycled and reused materials, an SDG 12 best practice that not only promotes responsible production and consumption through a circular economy model but also reduces greenhouse gas emissions and contributions to climate change.

A joint report published in February 2022 by the Department for Business, Energy and Industrial Strategy and the Newton-Picarte Fund notes that Chile is considered a stable economy, one that is characterized by open and competitive

[1] Kirchherr et al. (2017) and Camacho-Otero et al. (2018) describe evidence of this.

[2] Neptuno Pumps was the first company in the world to win the "Innovation of the Year" prize three times, as well as "Manufacturer of the Year" at the Pump Industry Awards in Oxford, England.

[3] This company is considered one of the 300 most sustainable companies in the world by the Global Opportunity Explorer in terms of the United Nations Sustainable Development Goals and was the first manufacturer in Latin America to be recognized as a distinguished company at the World Economic Forum, in Davos.

Table 1 Select development indicators: country profile (Chile)

	1990	2000	2010	2020	2021
World view:					
Population, total (million)	13.27	15.34	17.06	19.12	19.21
Population growth (annual %)	1.6	1.2	1.0	0.9	0.5
Poverty headcount ratio at $1.90 a day (2011 PPP) (% of population)	7.9	4.3	0.6	0.7	NA
People:					
Life expectancy at birth, total (years)	74	76	79	80	NA
Fertility rate, total (births per woman)	2.6	2.1	1.9	1.6	NA
Mortality rate, under-5 (per 1,000 live births)	19	11	9	7	NA
Environment, states, and markets:					
Annual freshwater withdrawals, total (% of internal resources)	2.3	3.5	4.0	4.0	NA
Urban population growth (annual %)	1.8	1.5	1.1	1.0	NA
Individuals using the Internet (% of population)	0.0	16.6	45.0	88.3	NA
Economy:					
GDP (current US$) (billions)	33.11	78.25	217.11	252.73	317.06
GDP growth (annual %)	3.3	5.0	5.9	-6.0	11.7
Inflation, GDP deflator (annual %)	22.5	4.6	8.9	8.8	4.5

Source: The authors, based on data from the World Development Indicators database (2022). Last Updated: 07/20/2022
Figures in blue refer to periods other than those specified, and NA indicates data not available. PPP refers to purchasing power parity

markets and a solid financial system. According to the World Bank, over the past few decades, economic growth in Chile has been almost consistently positive (see Table 1). This has been driven by the wider global commodity boom, which has been particularly good for Chile's main export, copper, although its other export sectors, including agricultural products, wood products, gold, and a nascent lithium sector, have also benefited. Despite the mining sector's size, most of the copper and lithium are exported in raw form. For this reason, the development of associated and knowledge-intensive activities has been described as a key priority for the sector (Department for Business, Energy, and Industrial Strategy, 2022).

According to the report, the Chilean government has recently been focusing on developing other sectors of the economy and positioning Chile as an excellent investment destination through the Invest Chile Agency. This agency has focused on encouraging technology development and innovation in several strategic areas, such as research and development (R&D), technological capacity, and human capital development. The Chilean strategy for the period 2018–2022 set out several plans for a considerable number of detailed initiatives linked to the creation of a dedicated Ministry for Science, investment in science and technology capacity, and some initiatives to bolster science and technology education and investment.

This case study illustrates the importance of both innovation and vision at Neptuno Pumps particularly in the period since 2014, when the company CEO, Petar Ostojic, embedded CE into the corporate strategy and, in doing so, made Neptuno Pumps a CE pioneer in Latin America thanks to a prominent presence in the media and communications arena. This fed into an extensive communication strategy that was implemented on social media and through presentations and webinars at companies, higher education institutions, and elsewhere. These wide-ranging actions led to Neptuno Pumps receiving several awards for its CE implementation from governments and international organizations such as the World Economic Forum and some United Nations agencies. Petar Ostojic was recognized as a CE promoter in Latin America by the World Economic Forum and awarded Highly Commended for Circular Economy Leadership by The Circulars 2017 in Davos.

2 Case Development

2.1 Sustainable Solutions and Innovation in Circular Business Models: The Context

It is known that the SDGs represent a fundamental step toward understanding the disadvantages of a traditional economic model, and certainly these goals inspire a commitment among civil society, governments, and the private sector to ensure the future of sustainable development worldwide. It is necessary to understand that integrating the SDGs into a business strategy helps to identify business and development opportunities for achieving the following main macro-objectives: eradicating extreme poverty, reducing inequalities, and tackling climate change. This type of combined action encourages the scaling up of projects to positive economic, social, and environmental effect (Galan-Ladero & Rivera, 2021).

In 2015, the Forum Empresa and EU-LAC Foundation registered a repository of the National Action Plans on corporate social responsibility (CSR) of the European Union (EU), Latin America, and the Caribbean. One of the contributions to come out of this initiative was the creation of working groups to share experiences and lessons learned among Europe, Latin America, and the Caribbean in order to establish a further repository of best practices in diverse areas. Five topics were identified for continuing this shared learning initiative (UNEP, 2022): (1) developing CE, protecting the environment, and generating clean energy, (2) identifying business and development opportunities in the context of the 2030 Agenda and the SDGs, (3) managing a socially and environmentally sustainable value chain, (4) reducing poverty and ensuring sustainable communities, and (5) promoting human rights and business worldwide.

CE is often associated with recycling, yet it also contributes to innovation, as the production, manufacturing, and selling of pumps made from recycled materials entail certain levels of creativity and transformation. This case study seeks to counter any criticism that has been levelled at the inclusion of CE in business strategies.

2.2 Efficiency, Innovation, and Customization in the Pump Industry: The Neptuno Case Study

Efficiency and innovation are two attributes that are embedded in Neptuno Pumps' DNA: for almost 50 years, the company has been developing a business model based on the design and production of energy-efficient equipment and the recycling and reuse of a wide range of resources, materials, and goods. However, in the last decade, the company began a radical transformation of its business and operational processes with a view to becoming a leader in the new CE paradigm.

The starting point was that for decades, owing to the reduced prices of materials and the lack of some industrial capabilities in the Latin American region, the mining industries had been operating with their own suppliers and parts and applying a traditional "throw-things-away" philosophy (see Fig. 1). However, the environmental and economic crises, coupled with an empowered consumer, have driven the industry to implement efficient, innovative, and sustainable and customized solutions in order to improve its productivity and mitigate the negative effects of climate change (Martinez-Alier, 2004). This is the why CE provides new opportunities for innovation while offering a perspective that goes beyond reuse (Stibbe et al., 2018).

This CE model calls for Neptuno Pumps to make and sell equipment or lease used equipment and then repair and resell it, which sets it apart from the traditional "use-and-discard" practices seen in mining. Like for new products, customers get a one-year guarantee when buying recycled pumps, yet these pumps are 60–70% of the cost of new ones and can be delivered in a much shorter time frame. Instead of extracting raw materials, the CE model proposes reusing old materials in new production processes, which not only extends the product's life cycle but protects the natural resources from scarcity (Word Pumps, 2016).

In 2016, Neptuno Pumps implemented innovative and efficient energy solutions based on its location—at the center of the world's biggest mining industry, in Chile. This initiative won two Pump Industry awards in the "Technical Innovation of the Year—Projects" category and other international innovation and sustainability awards. A relevant question that emerges here is: being more efficient is not enough; how can we learn from Neptuno Pumps? The steps followed in this case could be applied to other situations and contexts. They are the following:

Fig. 1 Some pump components under the "take-make-dispose" linear approach. Source: World Pumps (2016), courtesy of Neptuno Pumps

Table 2 Select results from Neptuno Pumps: period 2016–2018 (average) and current data

Period 2016–2018				Current data—2022[a]	
Annual sales	New jobs created	Energy saving up to	Waste reduction up to	Total revenues	Employees
+25%	+15%	70%	75%	$394,000	43

Source: The authors, based on data from Neptuno Pumps (2018) and RocketReach.co (2022)
[a]Available online at https://rocketreach.co/neptuno-pumps-profile_b5eaddc9f42e7af0 (accessed on 31 July 2022)

1. The first step aimed to manufacture using technology 4.0 and transform the business model into one based on innovation and CE. The strategic process begins in the design thinking phase and provides for the use of reused materials during the product definition stage. A vertical integrated model based on the company's design, manufacturing, and repairing activities facilitates implementation of this circular perspective. This model delivers a more efficient system by taking advantage of the technology used to design the products and the reverse logistics model already present in Neptuno's supply chain. Thus, by using 3D printing, finite element analysis, metal smelting simulation, and computational fluid dynamics software, it was possible to design various highly energy-efficient industrial solutions that utilized resources efficiently and competitively, thereby reducing manufacturing costs and greenhouse gas emissions. These results are demonstrated in Table 2.

2. The second step was to collect data in order to evaluate the economic, social, and environmental impacts that this CE model could have on industry in Chile in particular. In terms of environmental, social, and economic key performance indicators (KPIs), by remanufacturing, it is possible to reduce greenhouse gas emissions, avoid generating waste from materials, and lower production costs in comparison with those of traditional manufacturing processes. Previous research carried out by Neptuno Pumps on what happens at the end of the life cycle of its pumps identified that industry in Chile generated up to 400 tons of waste per month in 2018. Neptuno's new business model has enabled the creation of high-quality new jobs in a Chilean region that, since 2012, had been one of the three worst regions in the country for unemployment. The new business model has created positions for workers not only in Chile but also in countries such as Mexico and Peru (Kowszyk & Maher, 2018). In addition, the company rejigged its processes to make its pumps easier to disassemble and to provide access to the equipment's most expensive parts. Neptuno Pumps also works with customers to actively source material that can be recycled, and it hopes to build more local manufacturing capacity so that it can be closer to its suppliers. On an economic level, this model is of relevance for fast-growing emerging markets where development cash is scarce, and time and logistics capabilities are key factors in the delivery of heavy goods.

3. The third step implemented by Neptuno Pumps was to quantify the advantages of reusing material from worn-out pumping equipment in order to convince its managers, shareholders, and employees to adopt a CE approach. One of the challenges relating to implementation of a CE model that not everyone is entirely familiar with is competition with large multinationals: these companies are often more popular because of solid marketing campaigns. In addition, not all stakeholders (internal and external) believed that Neptuno could offer an effective CE model based on reverse logistics, repair, and remanufacturing (World Pumps, 2016).

First, to make the benefits tangible to shareholders and stakeholders, the company developed a few pilot projects and measured their results. Second, focusing on the internal stakeholders, the company started to involve different departments in this activity and took the time to work with them to define roles and responsibilities; thus, Neptuno changed the cultural organization surrounding the CE model and communicated to employees that they were all working together to achieve the same goal: to produce highly efficient goods.

The awards that Neptuno Pumps has won reinforce the idea that the company's implementation of a CE model has been a success story in the industry. Neptuno Pumps was the first company in the world to win the "Technical Innovation of the Year" prize three times (between 2013 and 2016); it was the Avonni 2014 Winner for Mining and Metallurgy and the "Manufacturer of the Year" at the Pump Industry Awards in 2017 (Oxford, England); it was a finalist in the "Environmental Contribution of the Year" awards in 2016; and it won the National Award for Companies of the College of Engineers of Chile 2016, among other achievements.

4. The fourth step was for Neptuno to use the measured actions and results to commit to reaching levels of 90% recycling. Furthermore, through the digitization of its products and services, Neptuno Pumps made it possible for clients to track and trace all processes, which ensures efficient use of resources and effective management of repairs and preventive maintenance. As reported by Neptuno Pumps, in 2015, a project was carried out in Antofagasta, Chile, to upgrade the pump systems with new, custom-engineered energy-efficient equipment. This reduced energy consumption by 15%, thus saving more than €500,000 in costs and avoiding 3798 tons of CO_2 emissions. Furthermore, the project allowed assets such as the vertical motors and other old pump components to be reused and material from old worn pumps to be recycled to produce the new equipment with a payback period of just one year, which had invaluable benefits for the local community. In 2016, the company's CE model enabled a particular client to save €120,000 by remanufacturing several pumps and to reduce energy consumption by 60%, waste by 75%, and carbon emissions by 70% by choosing to reuse equipment rather than buying new. As shown in Fig. 2, the CE approach is a restorative system by design: rather than extracting natural resources, used materials can be recovered and reused, thereby protecting natural resources from overexploitation.

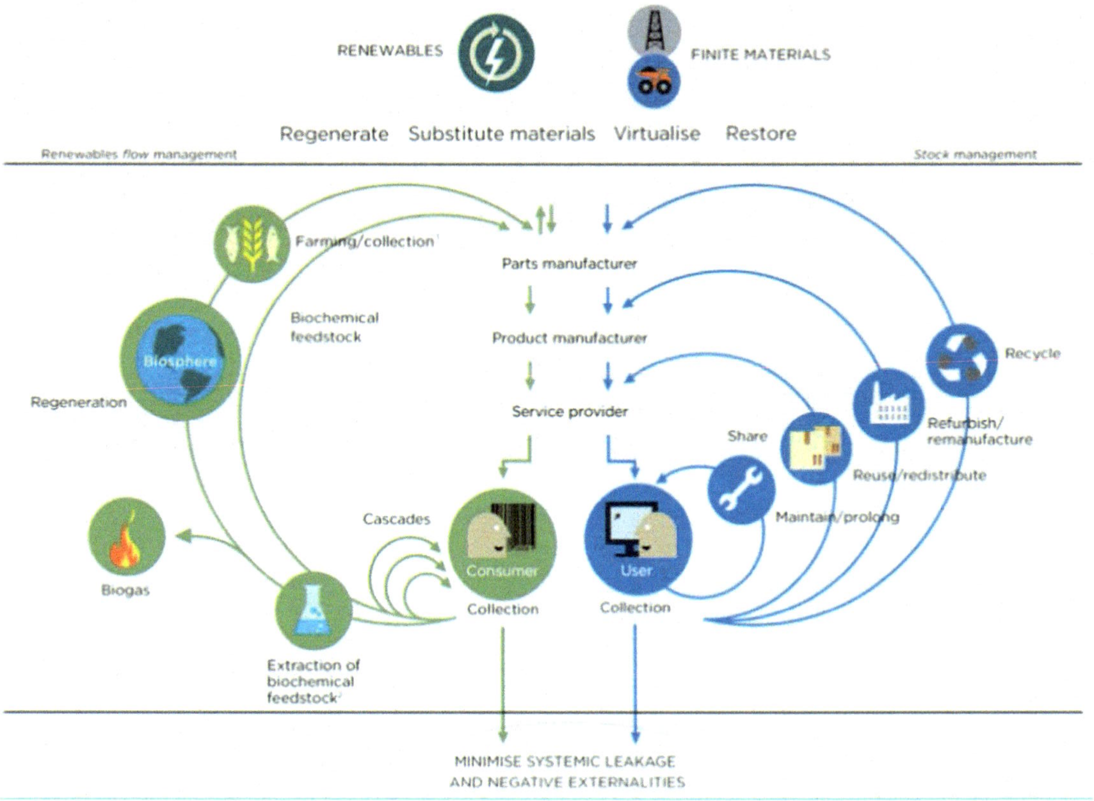

Fig. 2 CE diagram and the benefits generated in mining. Source: World Pumps (2016), courtesy of the Ellen MacArthur Foundation

5. Finally, Neptuno Pumps' CE model does not focus only on customer relations; it also aims to create partnerships with organizations relevant to the company. Communication is critical to standing out in this field and making the company an example for other organizations in the industry, and social media facilitates communication of the company's results and publications. In Chile, Neptuno Pumps works with the Center for Innovation and Circular Economy to deliver solutions that help businesses transition toward a CE model in Latin America. When it comes to the main results of this business model, communications and success stories shared on social networks have shown how much the CEO of Neptuno has inspired other leaders in the region by publicizing the benefits of CE in Latin America in particular. Neptuno was the first Chilean company to be selected as Runner-Up for the Circular Economy Enterprise Award and Finalist for the People's Choice Award at the World Economic Forum's The Circulars 2016 in Davos, precisely because of the wider impact of its work inspiring stakeholders' groups and the public in general to adopt CE principles (see Table A1 in Appendix 1).

Furthermore, pump functionality can be improved through better design and by seeking out new materials that maximize product life cycles. By testing and implementing new technologies and providing transparent information about the benefits of remanufacturing, it is possible to raise awareness among clients. And awareness of company supply chain practices and end consumer habits facilitates better consumption choices based on sustainable processes, which in turn creates more business based on environmental awareness and values.

To sum up, for all these reasons, Neptuno is currently considered one of the 300 most sustainable companies in the world. This case study describes Neptuno's success story and the tangible social, economic, and environmental impacts of its strategy (see Fig. 3).

3 Conclusions

This case study shows that customized, innovative, and efficient initiatives are key to achieving a broad range of sustainable and environmentally focused objectives. In 2018, the pump market accounted for €40 billion per year, consumed 10% of the energy generated annually on the planet, and transported up to 80% of the water used everyday around the world, emitting significant quantities of CO_2 in the process (European Commission, 2022). This case study illustrates how these resource consumption and emission figures can be reduced by using industry 4.0 technologies in combination with a CE strategy and how, by doing so, Neptuno Pumps has become a leader at a world level.

As a vital component of the SDGs, CE aims to facilitate the creation of new sustainable businesses, generate high-quality new jobs in the labor market, and fight climate change. That is exactly what Neptuno has achieved at international and national level since adopting a CE model, as demonstrated in this case study.

Fig. 3 Vertical turbine pumps made from reused parts in Chile. Source: Neptuno Pumps (2018)

For Neptuno Pumps, the implementation of an efficient and sustainable business model has made achievement of SDG 12 that bit more possible, as it has enabled the company to, inter alia, recycle hundreds of thousands of cubic meters of water that is wasted in other industries, reduce the annual carbon and CO_2 emissions from its manufacturing processes, and reuse thousands of tons of pump material that had previously been wasted. This case study demonstrates that sustainability, innovation, and client customization are the three pillars of a successful sustainable marketing initiative. For marketers, stakeholders, and shareholders, quantifying these types of initiative will ensure that socially responsible marketing programs and sustainable business models are valued for their positive effects on the community without detracting from the core objectives of these initiatives. Furthermore, the positive impacts of these initiatives become more evident through comparisons and by using KPIs for each of the three main aspects: social, economic, and environmental.

To sum up, the sustainable business model implemented by Neptuno Pumps has delivered high-quality jobs in a desert region that had been singled out as one of the three worst areas in the country for unemployment. Therefore, it is reasonable to believe that extending this business model will significantly increase jobs for qualified workers not only in Chile but in other countries such as Mexico and Peru, where Neptuno Pumps now has a strong presence.

Additionally, this case study also showed that when designing a value proposition for their CE-based sustainable business models, marketers and CEOs should place particular emphasis on disseminating a positive image of these models.

4 Discussion Questions

1. Analyze the situation of Neptuno Pumps as presented in this case study and identify the steps you think are necessary for setting up a systematic method of dealing with the problems described. Is your method the same as Neptuno's? Discuss.
2. Demonstrate how the SWOT matrix (i.e., strengths, opportunities, weaknesses, and threats) might be used to identify potential areas for development in a company like Neptuno Pumps.
3. Identify some of Neptuno Pumps' stakeholders and their interests. In your opinion, what potential conflicts of interest are there in this case study?
4. In your opinion, how can a company create value for its stakeholders? Can you identify two or three KPIs for each aspect (economic, social, and environmental) of the pump industry?

Appendices

Appendix 1

Table A1 Neptuno Pumps' circular economy prizes and awards (period 2013–2017)

Award	Category	Ranking	Year
Pump Industry Awards (PIA)	Manufacturer of the Year	Winner	2017
Pump Industry Awards (PIA)	Technical Innovation of the Year—projects	Finalist	2017
Pump Industry Awards (PIA)	Technical Innovation of the Year—projects	Finalist	2016
Pump Industry Awards (PIA)	Environmental Contribution of the Year	Finalist	2016
World Economic Forum—the Circulars Awards	Circular Economy Enterprise Award	Runner-up	2016
World Economic Forum—the Circulars Awards	People's Choice Award	Finalist	2016
College of Engineering of Chile	National Award	Winner	2016
Pump Industry Awards (PIA)	Technical Innovation of the Year—projects	Winner	2015
Pump Industry Awards (PIA)	Manufacturer of the Year	Finalist	2015
Pump Industry Awards (PIA)	Supplier of the Year	Finalist	2015
Pump Industry Awards (PIA)	Technical Innovation of the Year—products	Finalist	2015
Avonni Awards	Mining and Metallurgy	Winner	2014
Pump Industry Awards (PIA)	Technical Innovation of the Year—projects	Winner	2013
Pump Industry Awards (PIA)	Manufacturer of the Year	Finalist	2013
Pump Industry Awards (PIA)	Supplier of the Year	Finalist	2013
Pump Industry Awards (PIA)	Manufacturer of the Year	Finalist	2013
Pump Industry Awards (PIA)	Supplier of the Year	Finalist	2013

Source: The authors, based on data from World Pumps (2016) and Neptuno Pumps (2022)

Appendix 2: Teaching Notes for the Case

Synopsis

Neptuno Pumps is a company that produces pumps in Chile. It operates in the world's biggest mining industry, in the Atacama Desert. It applies a business model based on the design and production of energy-efficient equipment and the recycling and reuse of resources, materials, and goods. Over the past ten years, it has been steadily shifting over to the circular economy paradigm, transforming both its business and operational processes in order to deliver sustainable and innovative services. The company uses industry 4.0 technological solutions, including three-dimensional printing, computer simulations, and other technological advances and recycled, remanufactured, and reused materials to provide fully customizable products and services that are able to fulfill each consumer's needs. This circular economy model allows the company to have a positive impact not only on the environmental aspects of the pump industry—by reducing demand for natural resources—but on the economic and social aspects too, by providing new local jobs and creating social value.

Potential Audience and Instructor's Material

This case study has been developed for use in "Marketing Management" courses and is appropriate for MBA and Executive Development Programs, as well as advanced undergraduate courses. The case study is appropriate for courses that deal with socially responsible marketing and even for specialized modules on sustainable marketing communications and environmental analysis.

This case study provides a detailed description of the circular economy model that was implemented in Chile by Neptuno Pumps: a producer of pumps that delivers energy-efficient equipment and sustainable services to consumers.

In executive programs, discussions about the case study may lead on to more focused discussions on sustainable management and issues related to the stakeholder value approach, circular economy models, and social value creation, among other topics.

Learning Objectives

1. To identify responsible consumption and production trends in a sustainable business model in Chile
2. To illustrate the importance of the efficient and client-customized use of natural resources in one of the most arid places on the planet, the Atacama Desert
3. To explore what marketers should focus on when designing sustainable business models that promote the circular economy (CE) in several regions and places
4. To describe a relevant initiative in the pump industry that delivers innovative, sustainable, and efficient solutions through recycling and reuse

Time Frame for Class Discussion

This case can be taught in one session of approximately 75 min.

Suggested Discussion Questions

1. Analyze the situation of Neptuno Pumps as presented in this case study and identify the steps you think are necessary for setting up a systematic method of dealing with the problems described. Is your method the same as Neptuno's? Discuss.
2. Demonstrate how the SWOT matrix (i.e., strengths, opportunities, weaknesses, and threats) might be used to identify potential areas for development in a company like Neptuno Pumps.
3. Identify some of Neptuno Pumps' stakeholders and their interests. In your opinion, what potential conflicts of interest are there in this case study?
4. In your opinion, how can a company create value for its stakeholders? Can you identify two or three KPIs for each aspect (economic, social, and environmental) of the pump industry?

Analysis

Answer to Question 1. *Analyze the situation of Neptuno Pumps as presented in this case study and identify the steps you think are necessary for setting up a systematic method of dealing with the problems described. Is your method the same as Neptuno's? Discuss.*

In a socioeconomic context where so much attention is focused on climate change and the Sustainable Development Goals, consumers expect companies and organizations to reflect these issues in their conduct and performance. As a result, companies have started to adopt sustainability-based business models, thereby ensuring, and sometimes even boosting, their chances of survival.

The circular economy allows companies, countries, and societies to satisfy their consumption needs while protecting the environment from human overconsumption. When adopting a circular economy model, companies also need to develop a process for delivering organizational culture change or to start directly with that process. Companies should consider the following:

1. The circular economy model that best suits them, taking into consideration:
   (a) Challenges
   (b) Functioning
   (c) Objectives and environmental, economic, and social KPIs
   (d) Specific actions for achieving the objectives
2. The product and the production process, taking into consideration:
   (a) The supply chain before and after the circular economy model is applied
   (b) The value chain before and after the circular economy model is applied
3. The stakeholders and shareholders (specifically the clients and the local community), taking into consideration:
   (a) Their demands
   (b) Their needs
   (c) Their satisfaction

Based on the above, companies' approaches to implementing circular economy models may include different phases, for example, the following:

1. Analyzing the current situation. A SWOT matrix may be useful for this.
2. Identifying how to deal with weakness and threats. A set of strategic questions may be useful for this. (For example, which circular economy model suits the company better? Is a cultural change required in the company? Is governmental support needed? In your opinion, what is the role of the mass media?)
3. Finding providers. The Porter analysis may be useful for this.
4. Measuring results. To do this, specific KPIs need to be set.
5. Communicating to stakeholders using the most suitable communication methods (mass media, intranet, newspapers, etc.).

Answer to Question 2. *Demonstrate how the SWOT matrix (i.e., strengths, opportunities, weaknesses, and threats) might be used to identify potential areas for development in a company like Neptuno Pumps.*

To identify potential areas for development in a company like Neptuno Pumps, a good starting point is to analyze the SWOT matrix to determine company strengths, opportunities, weaknesses, and threats.

Neptuno Pumps has several strengths. It applies an innovative and sustainable business model that allows it to generate 60% of its goods using recycled materials and reusing waste equipment. This circular economy model has enabled the company to deliver products that are up to 30% cheaper, 25% more energy-efficient, and 90% of which are made with recycled and reused materials. Neptuno Pumps is an innovative company that has a strong media presence and whose sustainable efforts have been recognized by different awards bodies around the world.

Among its opportunities is the fact that it is located in Chile, which is considered one of the most stable economies in the region. That stability not only drives development in other sectors of the economy but makes the country one of the world's top investment destinations. Consumers' greater sensitivity to environmental issues also provides new opportunities for companies that implement circular and sustainable business models to provide value to their stakeholders.

Potential weaknesses are related to the criticism surrounding circular economy models. As Corvellec et al. (2022) indicate, the circular economy lacks clear theoretical grounds and limits, and its implementation deals with structural obstacles. These models are usually seen as purely focused on recycling, which may be considered a weakness of the model and hence a weakness for Neptuno Pumps to overcome, because it needs to clarify and convince critics that the circular economy encompasses the entire production system, from the input materials to the final product.

Among the threats is the fact that for decades the mining industry operated according to a "throw-things-away" model. Although climate change and other socioeconomic crises have now highlighted the need for more sustainable solutions, the inertia of the industry may be seen as a threat because it may require longer to adopt the necessary changes. And these changes must be adopted in the short term not only to stop but to reduce the negative effects of worldwide overconsumption on the climate and our societies.

Answer to Question 3. *Identify some of Neptuno Pumps' stakeholders and their interests. In your opinion, what potential conflicts of interest are there in this case study?*

The aim of any organization is to create value for consumers and stakeholders in general and in so doing to satisfy their needs and wants. This in turn requires that the firm/organization/institution first identifies the various specific interests of diverse groups.

From the perspective of institutions and government, the application of a circular economy model may set an example for other businesses or industries, which would therefore create value for other segments of society. But, at the same time, this may be seen as a threat in certain circles that depend on business models based on the "throw-things-away" system and thus on the production and sale of new items each cycle.

The company itself receives awards and distinctions worldwide, has a strong media presence, and is considered to be among the 300 most sustainable companies in the world. At the same time, all these achievements require energy, research and innovation, and a strong organizational culture geared toward the circular economy.

Competitors can use Neptuno Pumps as a benchmark, as can other companies whether in the same sector or not. This shared learning may strengthen the sector, although some parts of it may have difficulties adopting new business models or bringing about cultural change in their production processes.

The industry itself may also benefit from a circular economy model because it requires fewer input resources (materials), and hence there would be more goods available for use elsewhere. At the same time, a change of model may meet resistance from within the sector.

For clients, the availability of products and solutions is maintained while prices are cheaper, and quality and efficiency are higher. But acceptance of this also involves a change in mindset that not everyone is prepared for.

Society at large also benefits from the generalized adoption of circular economy models. Such models usually imply the creation of new jobs, a reduction in negative environmental impacts, a decrease in prices, less waste, etc. But they also require greater awareness of the global impact of individual actions and a commitment to considering the consequences of consumption.

Answer to Question 4. *In your opinion, how can a company create value for its stakeholders? Can you identify two or three KPIs for each aspect (economic, social, and environmental) of the pump industry?*

In a managerial sense, whether or not a company creates value for its stakeholders depends on its ability to provide goods or services that not only meet consumer needs but include added benefits that have importance for them, for example, production processes that deliver on consumer preferences, produce higher quality results, or offer better post-purchase services, among others.

Some of the ways in which Neptuno Pumps is creating value for its stakeholders through its circular economy model are the following:

- Providing clients with competitive products, thanks to the development of more efficient and sustainable solutions
- Reducing prices while maintaining or even increasing quality
- Reducing negative environmental impacts by reusing, recycling, and remanufacturing resources, materials, and goods
- Creating social value by creating new jobs
- Promoting the company's long-term commitment to sustainability

Neptuno Pumps have identified several KPIs to support this value creation. They include the following:

- Economic:
  Pumps that are 30% more efficient
  Pumps that are 30% cheaper
- Environmental:
  60% of goods made from recycled materials
  75% reduction in waste
  70% reduction in carbon footprint
  Products that are 25% more energy efficient
- Social:
  A 15% increase in new jobs
  New jobs in three countries (Chile, Mexico, Peru)

In general, having a circular economy business model can build and shape the reputation of a company and help give it a competitive advantage. For companies like Neptuno Pumps, this model may be thought of as how various stakeholders extract particular value, reward, utility, or benefit in exchange for their respective contributions to the business; that is, the development of a value proposition requires an approach to providing the required results through differentiation.

In this study, the real case of Neptuno Pumps, a pump producer, was used to illustrate the importance of the circular economy. This chapter suggests that implementing a CE model can provide value to all stakeholders, as long as there is an effective communication strategy in place to ensure society receives this value.

References

Akehurst, G., Afonso, C., & Gonçalves, H. (2012). Re-examining green purchase behaviour and the green consumer profile: New evidence. *Management Decision, 50*(5), 972–988.

Altinbasak-Farina, I., & Burnaz, S. (Eds.). (2019). *Ethics, social responsibility and sustainability in marketing*. Springer. https://doi.org/10.1007/978-981-13-7924-6

Bhattacharyya, J., Dash, M. J., Hewege, C., Makam, S. B., & Lim, W. M. (2021). *Social and sustainability marketing: A casebook for reaching your socially responsible consumers through marketing science*. Routledge.

Camacho-Otero, J., Boks, C., & Pettersen, I. N. (2018). Consumption in the circular economy: A literature review. *Sustainability, 10*(8), 2758.

Corvellec, H., Stowell, A. F., & Johansson, N. (2022). Critiques of the circular economy. *Journal of Industrial Ecology, 26*(2), 421–432.
Cuevas, J. R. (2016). Conciencia social y ecológica en el consumo. *Estudios Sociales. Revista de alimentación contemporánea y desarrollo regional, 25*(47), 301–322.
Department for Business, Energy & Industrial Strategy. (2022). *Partner country case study: Chile. Final evaluation of the Newton fund.* The Newton Fund. https://assets.publishing.service.gov.uk/government/uploads/system/uploads/attachment_data/file/1055509/newton-fund-evaluation-case-study-chile.pdf
European Commission. (2022). *The EU blue economy report. 2022.* Publications of the Office of the European Union.
Galan-Ladero, M. M., & Rivera, R. G. (Eds.). (2021). *Applied social marketing and quality of life. Case studies from an international perspective* (Book series 'Applying quality of life research: Best practices'). Springer. https://doi.org/10.1007/978-3-030-83286-5
Kirchherr, J., Reike, D., & Hekkert, M. (2017). Conceptualizing the circular economy: An analysis of 114 definitions. *Resources, Conservation and Recycling, 127*, 221–232.
Kowszyk, Y. (Director), & Maher, R. (2018). *Case studies on circular economy models and integration of sustainable development goals in business strategies in the EU and LAC.* EU-LAC Foundation.
Lubowiecki-Vikuk, A., Dąbrowska, A., & Machnik, A. (2021). Responsible consumer and lifestyle: Sustainability insights. *Sustainable Production and Consumption, 25*(January), 91–101.
Martínez-Alier, J. (2004). Los conflictos ecológico-distributivos y los indicadores de sustentabilidad. *Revibec: Revista de la Red Iberoamericana de Economía Ecológica, 1*, 24–34.
Stibbe, D. T., Reid, S., & Gilbert, J. (2018). *Maximising the impact of partnerships for the SDGs. A practical guide to partnership value creation.* The partnering initiative and United Nations Department of Economic and Social Affairs (UN DESA). Accessed 16 May 2022.
United Nations Environment Programme (UNEP, 2022). *Sustainable lifestyles. Empowering people to live better and lighter.* https://www.unep.org/explore-topics/resource-efficiency/what-we-do/sustainable-lifestyles
World Pumps: The International Magazine for Pump Users (2016). *Energy efficiency: The drive for improved industry benefits.* No. 587, January.

Part IV

Sustainable Multinational Enterprises

YPF Sustainability: Renovation of Service Stations and YPF Full Stores

Enrique Carlos Bianchi and Cecilia Bianchi

Learning Objectives

1. Understand how companies in the energy sector can commit themselves to sustainability through programs and investments aimed at caring for the environment by minimizing environmental and social impacts.
2. Understand that corporate investments in sustainability have a positive impact on brand reputation and customer validation.
3. Discuss the advantages and disadvantages of different approaches to social marketing: technological, informational, educational, and behavioral, aimed at helping consumers save energy and water.

1 Introduction

YPF celebrated its first 100 years.[1] It was founded on June 3, 1922, under Hipólito Yrigoyen's administration, by the name of Yacimientos Petrolíferos Fiscales (YPF). Enrique Carlos Alberto Mosconi, an Argentinian military engineer best known as the pioneer and organizer of petroleum exploration and extraction in Argentina, was appointed as its general director by President Marcelo Torcuato de Alvear.

The incipient YPF—driven by its pioneering spirit in the exploration, production, and refining of oil—needed to generate minimum conditions that did not exist in Argentine Patagonia. Thus, it built houses, roads, schools, grocery stores, hospitals,

[1] Institutional video of YPF's 100th anniversary: https://www.youtube.com/watch?v=6MUsRCx5nn8

E. C. Bianchi (✉)
Universidad Nacional de Córdoba, Córdoba, Argentina
e-mail: enrique.carlos.bianchi@unc.edu.ar

C. Bianchi
Universidad Nacional de Jujuy, Jujuy, Argentina
e-mail: ceciliabianchi@fca.unju.edu.ar

E. C. Bianchi et al. (eds.), *Responsible Consumption and Sustainability*, Springer Business Cases, https://doi.org/10.1007/978-3-031-30742-3_11

Table 1 YPF leadership in Argentinian market

Market share	Out of every 10 liters of gasoline and diesel, 5 come from their pumps (51.6% market share)
Top of mind	7 of every 10 persons (spontaneous renown)
Loyalty card	"ServiClub" card has a record figure of 2,400,000 active members
Stores	YPF FULL in 1500 service stations

Source: Authors with data taken from the Surtidores (2018)

clubs, and ports and generated the necessary conditions for the settlement of workers in such inhospitable areas; it was even a pioneer in the foundation of many cities in that region.[2] Back then, not only did it produce the energy that Argentinian people needed, but it was also synonymous with economic development and Argentine presence, granting the country's sovereignty.

Having been one of the first integrated state oil and gas companies in the world, YPF is now the most important company in Argentina, with the greatest transformational impact, and—by far—the largest investor. It has always been a pioneer in the energy industry, and that spirit has been kept over the years in the strategic vision of supplying energy to the country and exporting it to the world.

Currently, it is the market leader as it gained a 60% fuel market share in 2022, the greatest in the last 3 years (Table 1).

Inspired by its founding pioneering spirit, for a decade, it has been leading the development of "Vaca Muerta," the second largest deposit of unconventional gas resources and the fourth largest oil field in the world. Today, it seeks to increase hydrocarbon production in a sustainable manner and energy self-sufficiency in order to become a future reliable energy supplier to the world.

YPF is consolidating *new business units*:

- *YPF light*, which promotes participation in the Argentine electricity generation market, has sustainably grown to become one of the main electric-power generators in the country and the second largest producer of renewable energy.
- *YPF Litio*, which advances in the identification of opportunities for the exploration, production, and industrialization of lithium, takes advantage of the fact that Argentina has around 25% of the world's reserves.
- *YPF Tecnología* is the research and development unit that leads the H2ar,[3] a collaborative consortium created to study the potential to produce, market, and export hydrogen.

[2]It led the foundation and development of numerous cities, such as Comodoro Rivadavia, Las Heras, Cañadón Seco, Caleta Olivia, or Plaza Huincul in Chubut, Santa Cruz, and Neuquén provinces, in its operation surroundings.

[3]The H2ar is a collaborative workspace between companies that already work or are interested in taking part of the hydrogen economy in the country; the consortium structured its value chain work through eight cells, from production to application.

- *YPF Agro* is the unit with a broad portfolio of products and services, including seeds, crop protection and nutrition products, silo bags, fuels, lubricants, and the technical advice needed by the agro-industrial area.

YPF has been the protagonist of Argentine development, by exploring the soil and the sea; producing, refining, industrializing, and marketing gas and petroleum products; and paving the way for future challenges.

YPF has a great future ahead. It aims at doubling oil production and significantly increasing natural gas production, by highlighting Argentina's vast resources, in 5 years' time.

Active management of energy transition is one of the fundamental pillars of the company's business strategy, which contributes to the reduction of its own emissions and the carbon footprint of the value chain and increasingly integrates renewable energies to its portfolio.

2 YPF and Its Commitment to Sustainability

Our global population is on the rise, and this is placing increasing pressure on our natural resources. In today's society, people use a wide range of resources (from oils, gas, and minerals to energy, food, and water), and the overuse of these resources contributes to unsustainable consumption and production. As they are vital to societal and economic development, it is critical that in economic development, the environment is not harmed as a consequence. Therefore, SDG 12 intends to urge sustainable economic growth, especially in developing countries, while it simultaneously prevents the degradation of ecosystems.

The SDG 12, one of the 17 Sustainable Development Goals (SDGs) established by the United Nations in 2015, aims for all consumption and production patterns to be sustainable by 2030. SDG 12 is meant to ensure good use of resources, improving energy efficiency and sustainable infrastructure, providing access to basic services and green and decent jobs, and ensuring a better quality of life for all. This is because current patterns of consumption and production are unsustainable and are contributing to climate change, biodiversity loss, and pollution. These are all important issues that need to be addressed and prevented.

YPF understands sustainability, in its three dimensions (economic, environmental, and social), as a way of creating value for shareholders, other stakeholders, and the nation as a whole. The priorities of its sustainability policy are integrated into the business strategy and are aimed at conducting responsible operations and strengthening YPF's long-term competitiveness, based on three pillars and four main axes (see Table 2).

> To YPF, sustainability is an integral part of who we are as a company and the way we run business. We are committed to our mission of producing and providing sustainable energy in a profitable, competitive and responsible way, creating value for our shareholders,

Table 2 YPF sustainability model

Pillars			
Corporate governance and transparency	Alliances	Innovation and technology	
Main axes			
Sustainable production	Climate action	Shared social value	People
• Operational eco-efficiency • Circular economy and waste • Low-carbon products	• Reduction of CO_2-CH_4 emissions • Cleaner energy • Adaptation to climate change	• Relationship with communities • Supply chain • Education in energy	• Diversity • Development of human capital • Health and safety

Source: Own elaboration, out of information taken from YPF Sustainability Report (2021)

customers, employees, suppliers, the communities where we operate and the country, states Carlos Alfonsi, Executive Vice President of Downstream YPF.

YPF's commitment is to produce energy which is important to citizens in their daily lives, to cook and switch on lights, the same energy that allows hospitals to operate, streets to be lit, and factories to produce.

YPF Is Energy in Evolution It works to generate efficient and reliable energy through the development and production of conventional and unconventional oil and gas and renewable sources such as wind, sun, land, and water. In three industrial complexes located in La Plata, Luján de Cuyo, and Plaza Huincul, it generates fuels, petrochemical products, and lubricants, providing a complete range of products with a strong commercial presence in retail, agriculture, industry, and liquefied petroleum gas (LPG).

The supply of electricity grows with the construction of new thermal generation plants and the search for new renewable energy sources, in the future: 30 wind turbines from Manantiales Behr Wind Farm will distribute 100 MW to the Argentine Interconnection System, the equivalent of electricity used by 130,000 households.

YPF Is a Future That Challenges With the productive development of the "Vaca Muerta" geological formation, it is the protagonist of the unconventional hydrocarbon production in Latin America. YPF is taking its bets on the geographical and technological diversity of clean energy with alternative source projects, such as solar and biomass.

YPF Is Excellence in Products (Goods and Services) Leader in the fuel market, its products guarantee performance, adequate protection for each engine design, sustainability, and environmental care. The company is at the forefront of technological progress, and due to its service reliability, the quality of its fuels matches its industrial development and customers' demands.

YPF Is Everyday Closeness It works to satisfy customers' needs at each meeting point with greater flexibility and speed and the best operation service in its stations. With a logistics network of 1500 trucks, it manages to supply more than 1654 service stations throughout the country, making it the greatest coverage station in Argentina. To be close, YPF is building the service station of the future.[4]

3 Remodeling of YPF Service Stations and YPF Full Stores

The modernization plan of YPF service stations includes a meeting point, for meetings, recreation, product pickup, and energy supply, for travelers and customers called YPF Full Store, in more than 300 points in the country.

The initiative is part of a global trend, where outlets are increasingly focusing on the retail segment. Service stations "are not only the point of contact with our customers, but they are also our main presentation letter and communication channel with the community," says Leandro Caruso, Chief Marketing Officer of YPF.

Sustainability plays a major role since these new stations seek mainly to make rational use of water and energy.

YPF Service Stations have smart lighting control systems as well as sun protection techniques on glazed carpentry, which reduce the use of air conditioning equipment. They also have solar collectors to heat water. From the very beginning, within the transformation plan of its service stations, YPF has a strong commitment to environment care.

New YPF Full Stores respond to a renovation, an image associated with sustainability, that seeks to improve customer experience. It will add waste bins for garbage separation at source, high-speed Wi-Fi internet, and terminals to charge mobile devices at all tables. The New YPF Full is born with a sustainable, technological, and healthy concept.[5]

This was achieved with the financial assistance of Banco Nación which allocated US$25 million to the transformation of 1,347 service stations throughout the country, by an agreement signed with the YPF Operators Association. Flagship partners could have access to a series of financial instruments and special benefits to remodel and modernize all facilities and to carry out new projects.

[4] YPF #FutureServiceStation https://www.youtube.com/watch?v=4GWC5FpPvUg—Estudio GEMA Nueva imagen Full YPF Da Costa e Hijos S.A. https://www.youtube.com/watch?v=18xI26Tdyd8

[5] New YPF FULL https://www.youtube.com/watch?v=5BQr4qfyAG0—YPF the future is at #FULL. https://www.youtube.com/watch?v=AFPA40HgBzM

4 The New YPF Service Stations Protect the Environment

The brand-new YPF Service Stations have features that provide benefits to the environment in terms of energy savings, water savings, and waste separation, among other socioenvironmental issues. Some of the greatest impact initiatives include the following:

Building Infrastructure In building construction, a fast, resistant, and ecological system made up of concrete and mineralized wood fibers is used. A double layer of insulating protection is placed under the roof, and large openings have double glass to reduce thermal transfer and as acoustic insulation.

Construction includes the use of sustainable materials and high-performance technology: thermal materials, special glass resistant to shocks and sudden changes in temperature, solar panels and collectors, touchless flush toilets, and smart lights. In both material and function, they intend not only to be environmentally friendly but also to comply with the healthcare standards created as a consequence of COVID-19 pandemic.

These service stations also have photovoltaic cells that transform sunlight and convert it into electricity. Renewable energies release little carbon dioxide, reducing its harmful effect on the environment.

Besides, by using sustainable materials, they achieve maximum efficiency in their operations as they minimize expenses and costs and increase their profits and achievements. Their benefits are both economic and social as they protect the environment, show customer loyalty, and develop the surrounding community.

Energy Saving In their architecture, many service stations implement passive energy-saving measures by installing insulating materials in walls and ceilings. These are essential to reduce electricity consumption, especially in certain areas of the country that experience heat waves. In this case, for example, only one of the two inverter air conditioners will be used in summer at 26°C. On the other hand, other stations replaced efficient LED lights and installed photovoltaic solar panels.

All these new renewable facilities manage to generate electricity that is proportional to 45% of each service station consumption.

Water Recovery At some YPF service stations, one of their main innovations is rainwater recycling. Water is collected and then distributed in an independent circuit (separated from drinking water) to be used for toilets, parking cleaning, and garden irrigation and to save natural resources.

Water Saving in Toilets/Bathrooms Having comfortable and in good condition toilets/bathrooms for customers is a distinguishing characteristic of YPF Service Stations. The station located in Machagai, at 120 km from Resistencia, was one of the first to renovate completely its bathrooms so as to catch travelers on their way to and from Salta.[6]

It achieved this by investing in sanitary robotics (a technology made up of sensors, connected to the electrical network, and solenoid valves, normally closed) which allows the intelligent automation of sinks, urinals, toilets, and showers in changing rooms and restrooms for public use. Each one of the sanitary devices has a specific robotization kit. It is useful to reduce water and energy waste to zero, cut the chains of contagion (Covid), avoid lack of hygiene and comfort, and prevent vandalism.

In the past, if taps in public bathrooms had sensors on the spout, they only lasted a few days because people broke or stole them. Nowadays, these sensors are placed on the ceiling and are not visible. It is no longer necessary to share buttons or keys with other users; whenever the sensor detects any movement, such as showing of hands in the sinks and leaving urinals and toilets, it automatically lets out the flow of water.

Organic Waste Management and Reforestation YPF service station located in Misiones has its own orchard where vegetables and aromatic plants are grown. All organic waste from the service station is transformed to be used as fertilizer.

One hundred native species trees have been planted, together with the staff of the Amerian Portal del Iguazú Hotel, as social and environmental responsibility activity to reforest after the clearing caused by the new facility.

Dry Bath At the YPF service station on Buenos Aires—La Plata highway—a team from INTI's Sustainable Technology Program installed the first dry urinal that saves up to 100.000 liters of water and can produce nitrogen and phosphorus as fertilizer. The Sustainable Technology Program, together with society, aims to create sociotechnical responses in a sustainable way, promoting local development and protecting the environment. A dry toilet operates without flush water but takes advantage of the composting and desiccation process to decompose fecal matter. This type of toilet has a fermentation chamber to collect feces for long periods of time and a ventilation system to dry and ferment the organic matter. After at least 6 months, excreta are safe enough to be used as compost or dispersed in nature.

The urine drainage goes to the sewage system or to a storage container to be treated and to use its nitrogen and phosphorus as excellent fertilizers. Dry toilets have already been implemented in some airports of different countries in the world.

[6]Free Covid toilets in a Service Station: YPF Don Antonio, Machagai, Chaco. https://www.youtube.com/watch?v=Vutc5Xq659w

5 New YPF Full Stores

Today, YPF service station located in Caballito neighborhood in Buenos Aires not only has the latest technology fuel dispensers but also a new YPF FULL Store, aimed at customer's premium experience in terms of healthy gourmet food and products.

Miguel Ciancio, its owner, explains that this remodeling has to do with "making a more pleasant place based on the new trends in this type of cafeteria or retail business." With this change in trend, customers are coming on foot, just to go to the YPF FULL Store: "a large part of the people who enter our new Service Station come as pedestrians, without vehicles," he says.

With the new infrastructure, both for its furniture and for the lighting, it has become more comfortable and pleasant: "many customers who considered the Station as a 'home office' space now make good use of its high-speed internet, eat or drink something and work quietly in an agreeable place; they even hold virtual work meetings, instead of straying at home."

YPF believes that the new premises will become a friendly and engaging meeting point to customers and local people and that it will enable its staff to work more comfortably and with greater efficiency.

Cafeteria, Bakery, and Healthy Food Products As regards products for sale, both the quantity of food items offered and the quality of its bakery have been improved. Thus, in some YPF stores, a pilot test of premium pastries has already begun; it includes pastries from Maru Botana.[7] Besides, as food products are constantly being innovated, the station now offers healthier foods. Customers will find fresh fruits, salads, and natural juices on their new shelves.

Waste Segregation The new YPF FULL stores reinforce the concept of sustainability by adding specific baskets for waste segregation and by using ecological materials for cups and napkins, like poly-paper, and Full brand reusable bags. There is new in-store signage to educate consumers to behave responsibly.

Customer Experience The new infrastructure is part of an ambitious plan that seeks to provide customers with a full digital experience, personalized options, and an increasing range of available services.

With the YPF APP as an ally to integrate this customer experience, the new stations also feature a renewed architecture and design, with technologies applied to sustainability, and the transformation of service areas such as YPF FULL Stores and YPF service boxes.

In addition, some service stations will have a scanning system that reads the license plate of the entering car, allowing customers to be identified through ServiClub card so they will be offered promotions according to their tastes. The

[7]Maru Botana is a mother and a chef—https://marubotana.com/institucional/sobre_mi.php

incorporation and use of high-performance technological equipment make customer communication more direct and personalized.

6 Conclusions

With more than 100 years of experience, YPF continues to be an Argentine majority state-owned company that, with a pioneering spirit, today faces the challenges of sustainability in energy supply, being the protagonist of Argentine development. Sustainability priorities are incorporated into its business strategy, from exploration to fuel distribution through unconventional oil and gas production, creating value for its stakeholders.

The remodeling program of its service stations is not only aimed at providing a higher quality of service and experience to its customers but also at guaranteeing environmental care. By making different investments in construction (thermal materials) and in new technologies, it is able to make good use of sunlight (panels), recycle waste (in differentiated baskets), minimize water consumption (control sensors), save paper through digital payment methods (YPF Apps), and reduce groundwater contamination through wastewater treatment (dry toilets), among others. In addition, it demonstrates a commitment to customers' health by providing them with a variety of healthy food products (bakery, fruit, salads, natural juices, etc.).

7 Discussion Questions

1. What is YPF's commitment to sustainability?
2. What are the energy savings achieved by the new YPF service stations?
3. Does sanitary robotics investment imply an effective social marketing technology strategy to avoid wasting water? Is this solution valid when there is a low degree of consumer awareness?
4. What significant improvements in terms of customer commitment have the new YPF FULL stores incorporated?

Appendix: Teaching Notes for the Case

Synopsis

YPF celebrated its 100-year anniversary on June 3, 2022. Recently, more than 300 YPF service stations and their YPF Full stores from all over the country have been renovated, following environmental and social sustainability criteria. Today, they not only have state-of-the-art pumps but also an area to rest or relax and food stores where travelers and customers may enjoy a meal with healthy gourmet products.

This case study describes the remodeling program carried out by YPF which followed three sustainability criteria in the following: (a) construction, including the use of sustainable materials such as thermal material or special glasses resistant to shocks and sudden changes in temperature, (b) the use of modern technology in touchless activation devices and lights with motion sensors that save resources and help comply with health care (hygiene) regulations, and (c) investments in solar collectors to heat water and generate solar power to reduce energy costs.

Following the Sustainable Development Goal—SDG 12 (titled "responsible consumption and production")—the program demonstrates YPF's environmental care commitment and social commitment to its customers, assumed since it launched the renovation plan for its service station and FULL stores.

Potential Audience and Instructor's Material

The case is to be used in courses related to "marketing management," both at MBA and executive program level and in advanced courses at the bachelor's degree level. The case is appropriate to exemplify policies of energy companies that, committed to sustainability, seek to minimize economic, social, and environmental impacts.

By means of programs applied to investment, training, and education, they want to generate behavioral changes in consumers aimed at saving water and energy.

The case highlights the objective of the investments made also to maximize the customer experience.

Learning Objectives

1. Understand how companies in the energy sector can commit themselves to sustainability through programs and investments aimed at caring for the environment by minimizing environmental and social impacts.
2. Understand that corporate investments in sustainability have a positive impact on brand reputation and customer validation.
3. Discuss the advantages and disadvantages of different approaches to social marketing: technological, informational, educational, and behavioral, aimed at helping consumers save energy and water.

Time Frame for Class Discussion

This case can be taught in a 60–100-min session, depending on the number of students participating in the discussion activities and on the previous work done by the students.

Suggested Discussion Questions

1. What is YPF's commitment to sustainability?
2. What are the energy savings achieved by the new YPF service stations?
3. Does sanitary robotics investment imply an effective social marketing technology strategy to avoid wasting water? Is this solution valid when there is a low degree of consumer awareness?
4. What significant improvements in terms of customer commitment have the new YPF FULL stores incorporated?

Analysis

Answer to Question 1. *What is YPF's commitment to sustainability?*

This is an open question (there are no right or wrong answers). Students can give their personal opinion. Yet it is important that all students present strong arguments for their opinions. Some possible reflections may arise from reading the summary of YPF Sustainability Report[8] or its full version.[9]

On the other hand, the following ideas can be deduced from the case:

- The commitment to generate efficient and reliable energy.
- YPF Sustainability model, based on three pillars: corporate governance and transparency, alliances and innovation, and action, which are strongly related to sustainable production, climate action, shared social value, and people.

Answer to Question 2. *What are the energy savings achieved by the new YPF service stations?*

This is also an open question so each student can give a personal opinion to it. Yet, some of the possible inferences that may arise from the case are the following:

Actions: investments	Impact: energy reduction/saving
– Thermal construction materials – Intelligent lighting control system – Insulation protection on glazed carpentry – Double glass openings	– A decrease in the use of air conditioners in summer – Environment with more stable temperature throughout the year
– Solar collectors of hot water	– Reduction in the consumption of electrical energy

(continued)

[8] YPF Sustainability Report—Summary https://sustentabilidad.ypf.com/index_en.html

[9] Sustainability Report 2021—Full Version https://www.ypf.com/english/TheCompany/Documents/ypf-sustainability-report-2021.pdf?_ga=2.67122450.2038455726.1664567988-1090592056.1562474511

Actions: investments	Impact: energy reduction/saving
– Replacement of lighting for LED – Lights with motion sensors	– A decrease in energy consumption
– Photovoltaic solar panels	– An increase in the production of your own energy up to 45%

It is possible to discuss here whether or not YPF service stations should have carried out these technologically focused actions with educational and informative campaigns, by using tips like the following: "close the doors, air-conditioned environment"; "do not open the windows, air-conditioned environment"; "when leaving, turn off the lights"; etc.

Answer to Question 3. *Does sanitary robotics investment imply an effective social marketing technology strategy to avoid wasting water? Is this solution valid when there is a low degree of consumer awareness?*

Being an open question, each student can give their personal opinion. Here are some possible reflections that may arise:

Actions: investments	Impact: water reduction/saving
– Rainwater recovery system	– Save water to irrigate plants and feed the water tanks of the bathroom washing system – Save water to clean the service station car park
– Automatic tap opening sensors in sinks. – Intelligent automation of urinals, toilets – Automatic faucet closing sensors in the showers	– Save water due to less waste

Like in the previous question, it is possible to discuss here whether or not these technologically focused actions should have been accompanied by educational and informative campaigns, with tips such as the following: "take care of water when you shower, use only one card" or "remove your hands from the faucet, while rubbing them," among others.

You can discuss different consumer profiles with your students, for example, aware or unaware, and analyze which technological rather than informational social marketing approaches are more appropriate for those customer awareness profiles.

It can be a thought-provoking question. Students may analyze the theory of behavioral economics, which incorporates the study of psychology into the analysis of the decision-making behind an economic outcome, such as the factors leading up to a consumer buying one product instead of another. Nobel laureate Daniel Kahneman showed that our brain has two operating systems (one slow and one fast) for processing information about why people sometimes act against their own interests. He describes two distinct systems: System 1 is the brain's fast, automatic, intuitive approach which is highly susceptible to environmental influences; System 2 is the mind's slower, conscious, and analytical mode where reason dominates and takes into account explicit goals and intentions.

The economist Richard Thaler and legal scholar Cass Sunstein's Theory of Nudges can also be discussed. This theory is a concept in behavioral science, political theory, and behavioral economics that proposes positive reinforcement and indirect feedback as ways to influence behavior and individual or group decision making. A nudge makes an individual more likely to make a particular choice or behave in a particular way, by altering the environment so that automatic cognitive processes are triggered to favor the desired outcome.

One of the most frequently cited examples is the engraving of the image of a small fly on the center of the urinals of men's toilet at Amsterdam Airport Schiphol, by which men's aim material improved (they all tried to hit the fly) and the toilets became considerably cleaner and more hygienic. Another is to place small reminder notices, as we have pointed out in the answers to the previous questions.

Answer to Question 4. *What significant improvements in terms of customer commitment have the new YPF FULL stores incorporated?*

The aim of the remodeling of YPF service stations is to improve customer service experience. Here, once again, a strong investment in technology has been implemented (new YPF App, customer identification when arriving with their vehicle, etc.) as well as a renewed architecture and interior redesign of the YPF Full stores with more comfortable and spacious meeting areas and free WIFI connection so that the station could also be used as a "home office."

Likewise, with marketing campaigns and better promotions and healthy food offers that tempt customers, customers may win more points by using their loyalty card—ServiClub YPF.

Further Reading

Ámbito. (2021). *Banco Nación financiará obras en estaciones de servicio por $2.500 M.* 26 de octubre de 2021. Disponible en: https://www.ambito.com/edicion-impresa/banco-naci243n-financiar225-obras-estaciones-servicio-2500-m-n5305344

Dabbah, Augusto, F., et al. (2015). *Sistemas de saneamiento seco, baño seco* (1a edición ed.). Instituto Nacional de Tecnología Industrial (INTI). https://www.inti.gob.ar/publicaciones/descargac/505

EconoJournal. (2022). *La venta de combustibles se ubicó en abril un 12,6% por encima de la prepandemia y alcanzó el mejor registro de los últimos 10 años.* Martes 24 de mayo de 2022. Redaccion EconoJournal. https://econojournal.com.ar/2022/05/la-venta-de-combustibles-se-ubico-en-abril-un-126-por-encima-de-la-prepandemia-y-alcanzo-el-mejor-registro-de-los-ultimos-10-anos/

El Cronista. (2021). *Estación del Futuro', el nuevo concepto de YPF que cambia las estaciones de servicio.* 27 de octubre de 2021. https://www.cronista.com/brand-strategy/estacion-del-futuro-el-nuevo-concepto-de-ypf-que-cambia-las-estaciones-de-servicio/

Estudio GEMA. *Nueva imagen Full YPF Da Costa e Hijos S.A. - Florencio Varela.* https://www.youtube.com/watch?v=18xI26Tdyd8

La Nación. (2021). *YPF inauguró la primera estación del futuro en La Plata.* 27 de octubre de 2021. https://www.lanacion.com.ar/economia/ypf-inauguro-la-primera-estacion-del-futuro-en-la-plata-nid27102021/

Raies, E. (2017). *Una estación de servicios estrenará un mingitorio seco que ahorra 100 mil litros de agua.* Xelpaís. http://porelpais.com.ar/estrenan-mingitorio-seco/

Robotización de baños públicos - Domus Robótica Sanitaria. https://www.domusrobotica.com.ar/
Sanitarios COVID free en una Estación de Servicio: YPF Don Antonio, Machagai, Chaco. https://www.youtube.com/watch?v=Vutc5Xq659w
Surtidores. (2018). *Crece la preferencia de los clientes*. YPF alcanzó la máxima participación de mercado de los últimos 3 años. El Portal de las Estaciones de Servicio. November 7. https://surtidores.com.ar/ypf-alcanzo-la-maxima-participacion-de-mercado-de-los-ultimos-3-anos/
Surtidores. (2019). *Sustentables y solidarias con la comunidad: Así son las nuevas Estaciones de Servicio YPF*. El Portal de las Estaciones de Servicio. January 24. https://surtidores.com.ar/sustentables-y-solidarias-con-la-comunidad-asi-son-las-nuevas-estaciones-de-servicio-ypf/
Surtidores. (2020). *Experiencia exitosa. Utilizando criterios de sustentabilidad, Estaciones de Servicio ahorran la mitad del consumo eléctrico*. El Portal de las Estaciones de Servicio. August 24. https://surtidores.com.ar/utilizando-criterios-de-sustentabilidad-estaciones-de-servicio-ahorran-la-mitad-del-consumo-electrico/
Surtidores. (2021). *Una nueva experiencia para el cliente: YPF inauguró su primera Estación de Servicio con estructura de última generación*. El Portal de las Estaciones de Servicio. January 26. https://surtidores.com.ar/ypf-inauguro-su-primera-estacion-de-servicio-con-estructura-de-ultima-generacion/
Surtidores. (2022). *Antivandálicos. Una solución única para las Estaciones de Servicio que atraerá a los clientes*. El Portal de las Estaciones de Servicio. January 06. https://surtidores.com.ar/una-solucion-unica-para-las-estaciones-de-servicio-que-atraera-a-los-clientes/
Sustainable Development Goals. (2015). *United Nations*. Department of Economic and Social Affairs. Sustainable development. https://sdgs.un.org/goals

Videos

YPF #EstaciónDeFuturo. https://www.youtube.com/watch?v=4GWC5FpPvUg
YPF 100 años. https://www.youtube.com/watch?v=6MUsRCx5nn8
YPF APP. https://www.ypf.com/productosyservicios/APP-YPF/Paginas/home.html
YPF FULL Comida y Cafetería. https://www.ypf.com/productosyservicios/Paginas/YPF-Full.aspx
YPF Llegó el futuro a #FULL. https://www.youtube.com/watch?v=AFPA40HgBzM
YPF Nueva FULL. https://www.youtube.com/watch?v=5BQr4qfyAG0
YPF Reporte de Sustentabilidad. 2021. https://sustentabilidad.ypf.com/
YPF Sustainability Report. (2021). https://sustentabilidad.ypf.com/index_en.html

When Responsible Production and Consumption Matter: The Case of Danone

Alicia Izquierdo-Yusta, M. Dolores Méndez-Aparicio, Ana Isabel Jiménez-Zarco, and M. Pilar Martínez-Ruíz

Learning Objectives The learning objectives are as follows:

1. Understand how the macro- and microenvironment can influence an organization's sustainability strategy.
2. Promote stakeholder involvement in the achievement of sustainability objectives.
3. Describe what Danone has done well in terms of resource management to allow the company to adapt to the changing environment (best practices).
4. Search, store, and process information related to Danone as a practice in decision-making and independent learning.

1 Introduction

The Sustainable Development Goals (SDGs) of the United Nations' 2030 Agenda are a milestone in overcoming the limitations of the current socioeconomic model while also implying a joint commitment by all stakeholders (governments, civil society, and the private sector) to the future of economic development. For this reason, it is essential to work toward integrating them into business strategies. In addition to reducing poverty and lowering the impact of climate change, strategic business goals are also achieved, thus encouraging the further development of projects with economic and environmental benefits.

A. Izquierdo-Yusta (✉)
Faculty of Economics and Business, University of Burgos, Burgos, Spain
e-mail: aliciaiz@ubu.es

M. D. Méndez-Aparicio · A. I. Jiménez-Zarco
Faculty of Economics and Business, Open University of Catalonia, Barcelona, Spain
e-mail: mmendezap@uoc.edu; ajimenezz@uoc.edu

M. P. Martínez-Ruíz
Faculty of Economics and Business, University of Castilla la Mancha, Albacete, Spain
e-mail: mariapilar.martinez@uclm.es

E. C. Bianchi et al. (eds.), *Responsible Consumption and Sustainability*, Springer Business Cases, https://doi.org/10.1007/978-3-031-30742-3_12

The SDGs are also strongly related to the circular economy concept.

Among the SDGs, number 12 has a special link with the circular economy, as this new economic trend is restorative and regenerative by design. It seeks to redefine products and services to reduce waste through innovation while minimizing negative impacts. The circular economy requires us to rethink products and services using principles based on durability, renewability, and reusability, as well as the ability to repair, replace, upcycle, and refurbish materials and reduce their use.

Adopting a circular economy strategy helps companies to use their natural and energy resources more efficiently, reduce their greenhouse gas emissions, create more jobs, and meet the SDGs, [1] in particular SDG 12 on sustainable consumption and production. This is what led us to analyze the case of Danone and its commitment to sustainability.

Focusing on SDG 12 and its implementation at Danone, this objective requires innovation as well as the creation of a new development model based on technological and industrial partnerships that promote research into raw materials and the reuse of surpluses and waste from traditional processes. [2]

2 Case Development: Danone and the SDGs [3]

2.1 Danone Evolution

Danone was founded in 1919 when Isaac Carasso developed a new product that began to be sold at pharmacies called Yogurt Danone. [4] The innovative new product showed health benefits, especially for people with intestinal problems.

In 1929, Daniel Carasso (Isaac Carasso's son) joined the family business and setup in France, launching the brand under the name Société Parisienne du yoghourt Danone. In the 1940s, it began its expansion into the American market, choosing New York as a gateway, and increased its brand portfolio with new variants of the original product.

In 1973, the brand merged with Gervese. Throughout the 1980s, the group consolidated its position in the European market. Then, in the 1990s, BSN-Gervais Danone sought growth drivers further afield. Thus, countries in Easter Europe, Asia, and Latin America became potential and logical targets.

Nowadays, Danone holds onto its original values, its mission still being to "bring health through food to as many people as possible." However, sustainability is

[1] See Liu and Ramakrishana (2021).

[2] See, e.g., EnAlimentos (2020), Universidad Carlos III (2021), Ainia (2022), and LaVanguardia (2022).

[3] All the information shown here has been obtained from the official websites: https://www.danone.com and https://www.danone.es

[4] See Cruz et al. (2022).

starting to permeate the company, to the point that its frame of action has turned to "One Planet. One Health."

2.2 Vision: "One Planet. One Health"

Danone's current vision can be defined as follows: "*to bring health to our planet and to generations of people through our company and its ecosystems now and in the future.*" This idea was born in 1972 when Antoine Riboud, the first company chairman and CEO, said, "There is only one earth; we only live once."

In pursuit of this vision, and reflecting on how people's health is interconnected with the planet's, in 2017, Danone created a new brand logo and company signature: "One Planet. One Health." Inspired by the growing number of people who cared about where their food came from, how it was grown, how it arrived to them, and brands' social and environmental practices, "One Planet. One Health" began to drive the company's decisions and actions.

Through "One Planet. One Health," Danone promotes a new global entrepreneurial approach that, in partnership with nonprofit organizations such as Danone Foods-Grameen Bank, [5] leads to new, innovative product and process developments, as well as considering cultural, social, emotional, and physiological aspects related to dietary practices in every part of the world. To bring health through food to as many people as possible, the brand plays an active role in the "food revolution."

Care for the environment—and by extension care for the planet—is part of Danone's "One Planet" mission, which involves a set of specific, tangible objectives that Danone sets year after year and considers strategic. At COP21 in Paris in 2015, Danone presented its climate policy objectives in two phases: "to achieve a 50% reduction in emissions by 2030 and to reach zero net emissions by 2050." Danone's concern for health is now more relevant than ever. Over the past two decades, the company's portfolio has evolved to offer healthy products in all four of its divisions: fermented dairy and vegetable products, mineral waters, infant nutrition, and medical nutrition.

2.3 Mission: "Bringing Health Through Food to as Many People as Possible"

In line with its "One Planet. One Health" framework of action and B Corp ambitions, Danone defined *"Bringing health through food to as many people as possible" as its purpose or "raison d'être"* (Danone, 2022j). The social and environmental objectives that Danone adopted as a "Société à Mission" (i.e., its mission) are aligned with the United Nations' SDGs and cover four themes related to health, environment, socialness (with Danone's employees), and inclusiveness (referring to

[5] See, e.g., Avelino and Palacio (2021) and Di Tomasso (2022).

engaging with actors and communities within the company's broader value chain) (Danone, 2022j).

The mission objective is as follows: "impact people's health locally with a portfolio of healthier products, with brands encouraging better nutritional choices, and by promoting better dietary habits" (Danone, 2021, 2022j).

Danone also preserves and renews the planet's resources by supporting regenerative agriculture, protecting the water cycle, and strengthening the circular economy of packaging throughout its ecosystem to contribute to the fight against climate change.

Building on this mission, the brand will be renewed in order to find a balance between its purpose and performance. For a century, *Danone has been faithful to its original purpose, namely selling healthy, delicious products tailored to the nutritional needs and eating habits of its consumers. Essential dairy and plant-based products, water, early life nutrition, and medical nutrition—all of its ranges share the same goal: to offer innovative, top-quality products that meet the expectations of consumers at every stage of life. And while the company's business is about nourishing people and helping them take care of their health, it never forgets that it is also about helping them to live better lives and share their enjoyment with others. This is a new journey for Danone, one which builds upon its values and sets it up for success* (Danone, 2021, 2022j).

2.4 Danone's 2030 Goals [6]

Danone believes that people's health and the planet's health are interconnected and that both need to be cared for and fed. Accordingly, it has set its goals for 2030 in line with the "One Planet. One Health" frame of action, offering a perfect response to the challenges and opportunities of the ongoing food revolution. For that, the brand has integrated a set of nine global and long-term goals into its business model, which aligned with UN's SDG. [7]

According to its business model, [8] the objectives of its business model most closely related to SDG 12 are the following:

1. Offer superior food experiences and innovation: According to its reports, Danone "commits to the highest quality and food safety standards" (Danone, 2022a). To that end, the brand chooses natural and sustainably sourced ingredients as well as simple and traditional recipes and clean labels. Danone's strong innovation

[6] All the information shown here has been obtained from the official websites: https://www.danone.com and https://www.danone.es

[7] See Danone (2022a, 2022b).

[8] See: https://www.danone.com/about-danone/sustainable-value-creation/our-company-goals.html https://www.danone.com/content/dam/danone-corp/danone-com/investors/en-all-publications/2022/shareholdersmeetings/committeereportva0422.pdf

capabilities are key fundamentals to creating superior food experiences for people, as this will remain the first driver of healthier and more sustainable choices.
2. Deliver superior sustainable profitable growth: Danone's roadmap focuses on accelerating growth, maximizing efficiency, and allocating resources with discipline.
3. Be certified as a B Corp: As Danone notes, B Corp Certification is given to businesses that meet high standards of verified performance, accountability, and transparency with respect to factors including employee benefits, charitable giving, supply chain practices, and input materials.
4. Grow manifesto brands: Danone plans to turn its brands—for all its product categories—into "true manifesto brands" that show a real commitment to change through a healthy and caring planet.
5. Foster positive solutions for the planet, preserving and renewing the planet's resources: Noteworthy actions include (1) using natural and sustainable sources of ingredients, as well as protecting soil health through regenerative agricultural practices co-developed with partners (it will even amplify its ambitious water stewardship journey), and (2) fighting against climate change by implementing carbon-positive solutions and aiming to achieve carbon neutrality by 2050.

3 Danone: Responsible Consumption and Sustainability

As a B Corp company since 2016 (B Corp 2022b; Danone, 2022i), Danone is developing numerous initiatives around responsible consumption, including sustainable production. To achieve the nine global objectives of its business model that underpin its "One Planet. One Health" vision from a global and long-term approach, Danone has developed important lines of action that have guaranteed it a position as a company 100% aware of sustainability (IMF, 2022). The main initiatives are as follows:

- Promotion of a circular production and consumption model as *corporate culture*, including raising *public awareness and involving its employees*
- *Optimization of resources* in production by harnessing alternative energies, recycling water, reducing emissions, etc.
- *Recovery* of packaging and subsequent *recycling* of plastic
- Research into *renewable materials*
- *Eco-design* that facilitates the recycling of packaging
- *Last mile cost reduction*
- Promotion of *healthier*, more sustainable, and more inclusive eating and drinking *habits*
- *Reduced food waste*
- *Regenerative agriculture*
- Social and economic development through *education, sanitation, job creation, technology transfer, and improved agricultural production*

Let us look at these initiatives in detail:

Corporate Culture As a B Corp company, Danone aims to promote a responsible and sustainable production and consumption model that ensures a more sustainable food system that allows progress toward the decarbonization of the planet (Danone, 2022d). Presented during COP21 in Paris, Danone published its new climate strategy to cut its worldwide CO_2 emissions in half by 2030. This goal is already being achieved in many of its production plants. As seen in Spain in the dairy and water divisions, an organic reduction in CO_2 emissions of 30% and 45%, respectively, has been achieved between 2008 and 2015, and they aim to reach 0% emissions by 2050 (Danone, 2022h).

To achieve this goal, employee involvement is necessary. Through the creation of the **"One person, one voice"** *program*, since 2018, each of Danone's 100,000 employees is invited to actively participate and shape the future toward the Danone 2030 Goals. Through empowerment, innovation, and co-creation, the project's commitment is to extend and involve all employees under the same vision of the circular economy (WTWCo, 2021). In terms of *social awareness of responsible and sustainable consumption*, through the *RECONNECT project* (Danone, 2022h), Danone encourages consumers to connect with nature. Through a platform open to citizen participation, ideas will be collected to bring nature closer to the city. Through other channels, Danone also seeks to bring together relevant representatives of the entire food chain. One example was the Envifood Meeting Point (Madrid Meeting, 2016), which brought together more than a hundred of the most prominent representatives of this industry to discuss the importance of producers in the development of environmentally friendly policies.

Optimization of Resources in Production Processes Danone has reduced its *energy consumption* by almost 20% in the dairy sector and by more than 12% in the water business over the last 10 years, thanks to the use of 100% renewable electrical energy in all its sites, factories, R&D centers, head offices, and logistics centers. In terms of *water consumption*, thanks to research and innovation in manufacturing processes and water reuse, Danone has achieved industrial water savings of 30% and 55% in its dairy and water divisions, respectively.

Concerning Waste Danone expects to recover 100% of the waste it produces in different countries. This will be accomplished at its facilities in Spain, for example, by using sustainably sourced materials and promoting the recycling of waste produced at all plants. To meet this objective, the company has recently obtained the Saica Natur seal **"From Waste to Resources: Zero to Landfill,"** which guarantees that all its water plants in Spain are zero waste: Lanjarón in Andalusia, Sant Hilari Sacalm in Catalonia, and Sigüenza in Castilla-La Mancha. Specifically, the company recovers more than 98% of its waste and ensures that it is given a second life, preventing it from ending up in landfills and thus converting it into new resources. Danone is committed to making its entire water portfolio 100% recycled plastic (rPet) by 2025, a goal it recently achieved with Lanjarón. At the moment, Font Vella has two bottles made entirely from rPet, in 1.5 l and 50 cl formats for the restaurant sector. In addition, its second best-selling format, the 6.25 l carafe, already contains 15% rPet.

To reduce waste, it is necessary to *recover packaging* generated in the out-of-home consumption channel. This allows it to be given a second life by incorporating it as raw material for the creation of new packaging. In Spain, this initiative is being carried out together with the Trinijove Foundation and Ecoembes, where the online sales and home delivery platform will integrate the collection and subsequent recycling of plastic bottles delivered to its customers into its service. The management of this waste will be carried out through the *Renueva* container sorting plant (Renueva, 2022), a Spanish project that aims to increase recycling in the out-of-home waste market and, at the same time, to train and create jobs for people with disabilities or in situations of social exclusion. In this way, the company is taking a further step in its commitment to the circular economy, helping its consumers to recycle plastic bottles and reintroduce them back into its value chain in the form of recycled plastic in its new bottles. Danone's e-commerce platform has obtained Clean CO_2 certification, meaning that the online sales and home delivery service is carbon neutral (Danone, 2022f).

The Use of Renewable Bio-Based Materials Through an ongoing project in collaboration with Nestlé, PepsiCo and Origin Materials, Danone launched the first bottle with 75% bio-based materials in 2021 and will manufacture bottles with 100% bio-based materials by 2025 (Nestlé, 2018).

Eco-Design that Facilitates the Recycling of Packaging Danone is working on eco-design and on its commitment to giving a second life to resources for the production of all its bottles. For this reason, in addition to using 100% recyclable materials, the label has been removed from the bottles and the cap has been integrated to guarantee total recycling. All the profits obtained are allocated to projects to clean up natural spaces. To make 100% of its packaging recyclable, reusable, or compostable by 2025, Danone has redesigned the "Danacol" brand bottle following the principles of design for recycling, in order to ensure that it is easily recyclable. To this end, it has eliminated its PET label and replaced it with a relief on the bottle itself, thereby improving its recyclability and reducing 0.72 g of plastic per unit. Since its launch in November 2021, this has amounted to an annual saving of 130,000 kg of plastic (Alimarket, 2022).

Last Mile Cost Reduction In its desire to become increasingly "**more local and sustainable**" (Expansion, 2019), Danone has divided the world into 13 autonomous decision-making regions within this new local approach. This strategy enabled the company to increase its sales by 2.9% in 2018, to €24.7 billion, with an increase in its operating margin of 51 basis points, to 14.45% of sales. For example, Danone's Horizon Organic milk aims to be farm-to-table carbon positive by 2025. *Horizon*, ranked as the largest USDA-certified organic dairy brand in the world, believes it will be the first US dairy brand to reach this milestone. Doing so will require shaking up industry standards in emission-prone agriculture (Heaven32, 2020) by reducing the cost of the "last mile" of delivery. That is, optimizing the circuit from the farm to the transportation hub and from there to households.

Promotion of *healthier*, more sustainable, and more inclusive eating and drinking *habits*. Alpro has been recognized by B Lab (B Corp, 2022a) as one of the best companies for the world in the environment category, naming it Best for The World B Corps 2021 in Environment (B Corporation, 2022). Sustainability is part of Alpro, Danone's brand of beverages and plant-based alternatives to yogurt, which was launched more than 40 years ago with the mission to change the way the world eats, contributing to the global "**food revolution**"—a movement to encourage the adoption of healthier, more sustainable, and more inclusive eating and drinking habits.

Food Waste Reduction Danone fights climate change by reducing food waste. After announcing the introduction of the best-before date in May 2021, Danone aims for more than 90% of its products to carry the best-before date to reduce food waste (Financial Food, 2021). Danone has also incorporated the "**Dates with Sense: Look, Smell, Taste**" TGTG label promoted by Too Good To Go to help consumers check whether a product is still fit for consumption past that date using their senses.

Danone also collaborates in campaigns and actions to raise awareness and sensitivity among society and its own employees (Danone, 2022c) through the Waste Warrior Brand movement, a community of companies from different sectors that use their power to combat the food waste. Danone aims to reduce the food waste generated in its factories and production sites by 50% by 2030, in line with the objectives set out in SDG target 12.3 of the UN 2030 Agenda. Additionally, we see Danone pursuing responsible and sustainable consumption in its operations—in order to reduce the impact of food waste—through projects such as the Zero Impact Operations program. This project aims to reduce waste throughout the value chain, from the factories to the final product. In this way, waste is revalued and can be reintroduced as a resource for other uses.

Regenerative Agriculture Danone is committed to regenerative agriculture, working directly with family farms. The idea is to engage farmers in best practices and training in sustainable production. Its goal for 2025 is for 90% of its dairy supply to come from farms with generational renewal (Danone, 2022e). One noteworthy example is the initiative in France called "**One Planet Business for Biodiversity**" (OP2B). The aim of this initiative is for French farmers throughout the country to learn how to carry out 100% regenerative agriculture from an inter-sectoral approach, beyond their supply chain, by 2025 (Expoknews, 2022).

Social and Economic Development of Its Partners Among its noteworthy initiatives, Danone pursues objectives such as "social and economic development through education, sanitation, job creation, technology transfer, and improved agricultural production," corresponding to SDGs 7, 8, 12, 13, and 15. An example can be found in the *Frezna project* (Danone, 2022g), which addresses the three main aspects of Danone's strawberry production in Morocco: social, economic, and environmental.

4 Conclusions

Today, sustainability is a central issue for both companies and consumers, and society is becoming increasingly aware of the importance of preserving natural resources. For companies, this is increasingly reflected in the development of more sustainable production and marketing practices. On the consumer side, there is growing awareness, which is usually reflected in the selection of more environmentally friendly products and brands.

Danone is one brand to have adapted an important part of its organizational culture to the SDGs. Traditionally, the brand has been committed to health care; today, under the slogan "One Health. One Planet," it recognizes that people and the planet are interconnected, meaning people's health affects the planet's health.

The brand's objectives are both in the area of learning and processes, as well as in strengthening relationships with its customers through education and experience creation. Innovation is constant, so in addition to investing heavily in resources and training, it also collaborates directly with organizations, companies, and institutions with a solid reputation in the field of sustainability. Furthermore, Danone's ambition to become a Certified B Corp underlines its long-term commitment to creating and sharing sustainable value for all, in line with its dual economic and social agenda. In today's world, big companies and their brands are fundamentally challenged in terms of whose interests they really serve.

The main actions developed by the brand are established in each and every stage of the value chain, giving a perfect example of what a true circular production and consumption system is. In the area of production, the company is committed to the collection and recovery of packaging, the use of recycled materials, and the use of eco-design. Meanwhile, as regards production, the company is committed to production models that focus on energy saving and reducing its carbon footprint.

Finally, Danone is committed to training, both among its employees and its customers. Communication has become a powerful tool not only to educate and raise awareness but crucially to encourage sustainable use and consumption behaviors.

5 Discussion Questions

1. How can we assess the implementation of these strategies in companies (CSR, circular economy, sustainability)? Is this a passing fad? Is it an attempt to convey a different image? Are the messages that companies are sending out true?
2. What is your take on the "roadmap" used by Danone to implement its sustainability and corporate social responsibility strategy, taking its business model into consideration?

3. Do you think Danone has communicated its sustainability commitment to consumers? Analyze some of the examples used in the case, or use others.

Appendix: Teaching Notes for the Case

Synopsis

Commitment by all stakeholders to responsible production and consumption is transforming the until now predominant form of business, as well as political and financial activities and consumer decision-making. Companies must be aware that natural resources are limited and can no longer be misused. This means seeking new production models, redesigning products, reusing materials that were previously considered waste, and reducing consumption. Consumers, meanwhile, should change their consumption habits. This latter group is demanding social responsibility from companies and brands through their actions, which are in turn raising greater environmental awareness. Legal and political aspects are also embracing this new paradigm. Danone is one company that has initiated change to comply with the Sustainable Development Goals (SDGs). Here, we analyze how this process is unfolding and the milestones that have been achieved.

Potential Audience and Instructor's Material

This case has been developed for use in courses on marketing, corporate social responsibility (CSR), and the circular economy. There are two potential audiences for this chapter: undergraduate students pursuing degrees in business administration or marketing and graduate students pursuing MBAs and master's degrees in marketing, consumer social responsibility, innovation, and the circular economy.

Time Frame for Class Discussion

Estimated class time is 100 min. Students must come to the discussion having read the chapter and looked for additional information (if they consider it necessary). It would be interesting to work on the case in groups of three students.

Suggested Discussion Questions

1. How can we assess the implementation of these strategies in companies (CSR, circular economy, sustainability)? Is this a passing fad? Is it an attempt to convey a different image? Are the messages that companies are sending out true?

2. What is your take on the "roadmap" used by Danone to implement its sustainability and corporate social responsibility strategy, taking its business model into consideration?
3. Do you think Danone has communicated its sustainability commitment to consumers? Analyze some of the examples used in the case, or use others.

Analysis

Answer to Question 1. *How can we assess the implementation of these strategies in companies (CSR, circular economy, sustainability)? Is this a passing fad? Is it an attempt to convey a different image? Are the messages that companies are sending out true?*

This is an open-ended question (there are no right or wrong answers), so each student can give their personal opinion. What is key is that students make strong arguments for their opinions.

Some possible reflections might include the following: Are companies following through on their corporate social responsibility? Are the SDGs the solution to economic and social problems?

Answer to Question 2. *What is your take on the "roadmap" used by Danone to implement its sustainability and corporate social responsibility strategy, taking its business model into consideration?*

This is also an open-ended question.

Here are some ideas to work on this question:

- How has Danone been implementing this strategy based on its vision and mission?
- What are the pros and cons of this strategy? Students should try to link this strategy with the principles of the circular economy (designing ways to minimize waste and pollution, continuing to use materials, regenerating natural systems).
- It would also be interesting for students to take into account that Danone is a B Corp company. How has this influenced its strategy?

Answer to Question 3. *Do you think Danone has communicated its sustainability commitment to consumers? Analyze some of the examples used in the case, or use others.*

This is also an open-ended question.

- By analyzing some of the initiatives implemented by Danone in its 2030/2050 sustainability strategy, students should assess how the company has dealt with communication. They should take into account communication campaigns, press reports, etc. It would also be interesting to analyze whether consumers are aware of the actions undertaken by Danone.

– It would also be interesting to carry out an analysis of a specific case of how Danone carries out its corporate social responsibility policy. Is society aware of these actions?

References

Ainia. (2022). *Nuevos usos para los excedentes lácteos*. https://www.ainia.es/ainia-news/nuevos-usos-para-los-excedentes-de-suero-lacteo-en-la-industria-quesera/

Alimarket. (2022). *"Nuestro reto es construir valor, también con envases que a los consumidores les apetezca comprar" - Noticias de Envase en Alimarket, información económica sectorial.*

Avelino, C. T., & Palacio, J. R. S. (2021). Ética y Responsabilidad Social en las alianzas estratégicas. El caso de las alianzas entre entidades lucrativas y no lucrativas. *REVESCO: revista de estudios cooperativos, 137*, 142–160.

B Corp. (2022a). *Sobre B Lab*. https://www.bcorpspain.es/b-lab-es

B Corp. (2022b). *Directorio B Corps*. https://www.bcorpspain.es/directorio-b-corps

B Corporation. (2022). *Danone Nordics*. https://www.bcorporation.net/en-us/find-a-b-corp/company/danone-nordics

Cruz, N., Abernathy, G. A., Dichosa, A. E., & Kumar, A. (2022). The age of next-generation therapeutic-microbe discovery: Exploiting microbe-microbe and host-microbe interactions for disease prevention. *Infection and Immunity*, e00589–e00521.

Danone. (2021). *Mission committee report 2021*. https://www.danone.com/content/dam/danone-corp/danone-com/investors/en-all-publications/2022/shareholdersmeetings/committeereportva0422.pdf

Danone. (2022a). *Company Dashboard 2020 Financial and Extra – Financial Results*. https://www.danone.com/content/dam/danone-corp/danone-com/about-us-impact/sust_value_creation/DASHBOARD-DANONE-2020-PERFORMANCE-ON-2030-COMPANY-GOALS-2.pdf

Danone. (2022b). *Danone. Goals by 2030*. https://www.danone.com/about-danone/sustainable-value-creation/our-company-goals.html

Danone. (2022c). *Where do you want to work?* https://careers.danone.com/en-global/home.html

Danone. (2022d). *"Raison D'Être"*. https://www.danone.com/about-danone/sustainable-value-creation/danone-entreprise-a-mission.html

Danone. (2022e). *Trabajamos estrechamente con granjas familiares*. https://www.danoneespana.es/cuidamos-el-planeta/danone-promueve-agricultura-regenerativa.html

Danone. (2022f). *Font Vella en Casa obtiene la certificación de carbón neutral*. https://www.danoneespana.es/noticias/noticias-destacadas/Font_Vella_en_Casa_obtiene_la_certificacion_Carbon_Neutral.html

Danone. (2022g). *Regenerative agriculture*. https://regenerative-agriculture.danone.com/es/projects/proyecto-frezna/

Danone. (2022h). *Cero emisiones en 2050, principal objetivo de la nueva política de climática de Danone*. https://www.danoneespana.es/noticias/noticias-destacadas/Cero-emisiones-en-2050-principal-objetivo-de-la-nueva-politica-climatica-de-Danone.html

Danone. (2022i). *Danone B Corp*. https://www.danone.com/about-danone/sustainable-value-creation/BCorpAmbition.html

Danone. (2022j). *Danone mission*. https://www.danone.com/about-danone/sustainable-value-creation/danone-societe-a-mission.html

elEconomista. (2016). *Congreso Envifood Meeting Point 2016, los retos del sector de alimentación y bebidas*. https://www.eleconomista.es/economiaeAm/noticias/7641017/06/16/Congreso-Envifood-Meeting-Point-2016-los-retos-del-sector-de-alimentacion-y-bebidas.html

EnAlimentos. (2020). *Danone lanza yogur fabricado con residuos alimenticios.*

Expansión. (2019). *Danone será cada vez más local y sostenible*. https://www.expansion.com/empresas/distribucion/2019/05/09/5cd33242e5fdea8e548b45b6.html

ExpokNews. (2022). *Danone recibe título honorario por su sostenibilidad*. https://www.expoknews.com/danone-recibe-titulo-honorario-por-su-sostenibilidad/#:~:text=Danone%20dio%20a%20conocer%20que%20el%20n%C3%BAcleo%20de,corporaci%C3%B3n%20creada%20por%20una%20ley%20francesa%20de%202019

Financial Food. (2021). *Danone apuesta por un futuro sin fecha de caducidad para reducir el desperdicio*. https://financialfood.es/danone-apuesta-por-un-futuro-sin-fecha-de-caducidad-para-reducir-el-desperdicio/

Heaven32. (2020). *La leche orgánica Horizon de Danone pretende ser carbono positiva de la granja a la mesa para 2025*. https://www.heaven32.com/internacional/la-leche-organica-horizon-de-danone-pretende-ser-carbono-positiva-de-la-granja-a-la-mesa-para-2025/

IMF. (2022). *Danone, una empresa 100% concienciada con la sostenibilidad*. https://blogs.imf-formacion.com/blog/mba/danone-empresa-sostenible/

LaVanguardia. (2022). *Por qué nunca deberías desechar el líquido del yogur*. https://www.lavanguardia.com/comer/materia-prima/20191204/472040477656/beneficios-liquido-yogur-suero-o-desechar.html

Liu, L., & Ramakrishna, S. (Eds.). (2021). *An introduction to circular economy*. Springer.

Nestlé. (2018). *Danone, Nestlé Waters y Origin Materials dan la bienvenida a PepsiCo a la NaturALL Bottle Alliance*. https://www.nestle.com/media/news/naturall-bottle-alliance-welcomes-pepsico

Renueva. (2022). *Recycling*. http://ecosysteme.danone.com/projectslists/renueva/

Di Tomasso. (2022). *Conicet transfiere tecnología a Danone*. Sus yogures tendrán probióticos estudiados por el organismo. https://www.saraditomaso.com/2022/01/27/conicet-transfiere-tecnologia-a-danone-sus-yogures-tendran-probioticos-estudiados-por-el-organismo/

Universidad Carlos III. (2021). *REVOLUTION OF RAW MATERIALS, BIOTECHNOLOGY AND NEW MATERIALS*. https://www.uc3m.es/ss/Satellite/InnovacionEmprendimiento/en/TextoDosColumnas/1371307350725/Revolucion_de_las_materias_primas,_Biotecnologia_y_nuevos_materiales

WTWCo. (2021). *How employee insights power Danone's sustainable growth agenda*. https://www.wtwco.com/en-CM/Insights/2021/09/how-employee-insights-power-danones-sustainable-growth-agenda

Socially Responsible Actions for a Sustainable Life Model: The Case of the BBVA Financial Group in Mexico

Lucirene Rangel-Lyne, Ricardo Alberto Sánchez-Jaime, Ma. De Lourdes Cuellar-Hernández, and Dionicio Morales-Ramírez

Learning Objectives

1. Describe the convergence of corporate social responsibility and sustainability.
2. Expose the sustainable impact that promotes responsible actions through involved third parties.
3. Analyze the importance of responsible consumption through a high-impact company in the financial sector.
4. Identify the main socially responsible actions of a leading financial group in its branch.
5. Ponder about the importance of establishing key alliances with the implicated groups, so as to generate a real change with a higher impact regarding the topic of social responsibility for a sustainable lifestyle.

1 Introduction

In response to social, economic, and environmental problems, world organizations have determined common objectives in favor of a sustainable lifestyle. As a strategy for the achievement of more favorable conditions for all, corporate social responsibility (CSR) has been adopted by companies that are aware of the need for a positive change for the environment in which they operate, at the same time, that new requirements arise in the consumption market.

At the beginning, the company was an entity focused on the maximization of valuable resources for its investors, which is why the only responsibility for companies was being stable and profitable in economic terms. Over time, this vision

L. Rangel-Lyne (✉) · R. A. Sánchez-Jaime · M. De Lourdes Cuellar-Hernández · D. Morales-Ramírez
Universidad Autónoma de Tamaulipas – UAT, Tampico, Tamaulipas, Mexico
e-mail: lucirene.rangel@uat.edu.mx; rsanchez@utaltamira.edu.mx; mcuellar@utaltamira.edu.mx; dmorales@docentes.uat.edu.mx

E. C. Bianchi et al. (eds.), *Responsible Consumption and Sustainability*, Springer Business Cases, https://doi.org/10.1007/978-3-031-30742-3_13

changed toward a sustainable paradigm, for the benefit of society and not only its own interest (Sarro et al., 2007). Therefore, in the last few years, the CSR has become a direct actor in the world participation for taking care of global problems.

Having said that, CSR is a topic of international interest because it impacts the three dimensions of sustainability: economic, social, and environmental. In 2015, countries within the United Nations (UN) agreed to take actions for the current world situation, establishing in that way the 17 Sustainable Development Goals (SDGs). In this way, a comprehensive agenda was proposed within the business sector and the government of the participating countries, in order to address topics such as poverty, healthcare development, social inequality, and economic growth, all of this approached within the world climate change context (UN, 2022).

Among the actions by these companies that stand out, the generation of policies and programs focused on consolidating a sustainable model has been emphasized, this meaning the capacity of satisfying the current needs without compromising the resources of future generations. Regarding the change that global economic entities can produce, a group with considerable importance due to the economic impact on nations is the financial sector. Referring to individual and organizational consumption, financial corporations have the capacity of enhancing the consolidation of sustainable models with higher coverage, since these corporations constitute a segment that stands out on the incorporation of CSR practices; due to their influence on the economic system, these corporations intrinsically manifest, since they indirectly impact on social and environmental aspects (García & Márquez, 2021).

One of the most remarkable responses of this business sector in correspondence with the commitment to attend to the objectives established by the United Nations was the formation of the Finance Initiative of the United Nations Environmental Program (UNEP FI).

UNEP FI is an association of the United Nations (UN) with the global financial sector, with the purpose of mobilizing the private sector toward sustainable development. This association participates with over 450 banks, insurance companies, and investors with the goal of creating a financial sector that serves people and the planet through positive impacts. It intends to improve people's life quality, without compromising future generations' resources, while attending the acceleration of the development of sustainable finances (UNEP FI, 2022).

Among the countries with pending environmental issues, at the same time presenting social and economic gaps of a priority nature, Mexico can be identified. According to data from the Secretary of Energy in Mexico (SENER), this country is supplied with 82.87% of primary energy coming from hydrocarbons (SENER, 2018), being as well an emitter of greenhouse gasses, which contribute to the development of disasters related to global warming (NOTICIAS ONU, 2019). In addition to this, socioeconomic gaps exist in terms of schooling and gender equality which are accentuated with global warming. As stated by the INEGI (2020), with a whole population of 128.9 million people, only 33.6 million (from 3 to 29 years old) were enrolled in the 2019–2020 school year (62% of the total), from which 740,000 did not conclude said period, also facing a gender gap in terms of economic participation (a labor participation rate of 42.9 is maintained for women against 77.6 for men).

Likewise, an example of the active companies is BBVA bank in Mexico, since BBVA is part of the sustainable development strategy through the UNEP FI, allocating private resources for the participation in the commitments of the pact of the United Nations.

2 Case Development

BBVA Mexico is the largest financing group that operates on Mexican national territory, as they manage assets that represent 22.9% of banking in Mexico. BBVA Mexico was founded in the year 1932 in Mexico City; this company currently offers different financial services toward customers and companies, and among these services are mortgage credits, credit cards, pensions, insurance, and other financial services focused on the business sector. In the year 2021, BBVA Mexico received the recognition of "Socially Responsible Company" from the Centro Mexicano para Filantropía (CEMEFI) (Mexican Center for Philanthropy, in English), which has been ratified during the last 21 years.

This bank has also focused on attending topics of corporate social responsibility divided into three main axes, these being the environment, society, and economy, addressed onto accomplishing four main objectives:

(a) Generating trust by serving more and better their clients, with transparency and integrity, always offering products and services with the highest quality; (b) providing to the collaborators the best conditions for their integral development; (c) being financially solvent and offering attractive returns to the investors; and (d) supporting social welfare as a result of the business activity. BBVA received the recognition of "Best Digital Bank in Mexico 2019" in that year's World Finance Banking Awards ceremony, the above for the reason of the digital transformation process of the sector that BBVA is leading and the innovation on their products and digital services.

BBVA Mexico Group indicates clearly in its annual 2021 report that they have committed to contributing to the sustainability and development of a community with an inclusive growth; thus, three lines of action have been implemented, which are the reduction of inequality, promotion of entrepreneurship, and creation of opportunities for all through education and support of research and culture, which greatly contributes to its CSR. The section will show achievements, programs, and activities done in accordance with each CSR dimension in order to support the community in Mexico.

2.1 BBVA Mexico: Economic Dimension

According to CEMEFI (2022), the economic dimension of the CSR refers to the generation and distribution of added value among collaborators and shareholders, involving the production and distribution of useful and profitable goods and services for the community in which they are located. In this sense, the shareholders and collaborators within this financial group have been benefited by maintaining more than 13,400 ATMs throughout the country, a portfolio worth 1,291,449 million Mexican pesos, 1,716 bank branches, and 15.6 million digital clients, and then achieve a net profit of 65,502 million Mexican pesos (BBVA Mexico, 2022a, b, c). This contributed to consolidate itself as the most recommended bank by customers in Mexico, achieving an honorable mention at the Global SME Finance Forum 2021, as an organization that works to expand access to finance to small- and medium-sized companies; this bank operates a global network of more than 220 members that brings together financial institutions and technology companies, to share knowledge, stimulate innovation, and promote the growth of small- and medium-sized companies.

As for the services that BBVA Mexico offers to generate added value, there are business and government banking that offer specialized services to the different levels of government and other government agencies, credit focused on developers for construction purposes, consumer financing to the automotive industry, mergers and acquisitions services, insurances, micro-business credits, and private investment banking, to name a few. The said services allowed BBVA Mexico to consolidate its leadership position in the current portfolio with a market share of 23.7% and an increase in bank deposits of 10.6% per year, according to data by the Mexican National Banking and Stock Commission in 2021.

On the basis of this framework, BBVA Mexico in its economic dimension supports the commitment of five of its 15 material topics, which are responsible investment and financing, risk and crisis management, regulatory compliance, corporate governance, and the resilience of the business model.

Green bonds: Having established itself as a leading company in its field, the financial entity has implemented important efforts regarding social responsibility through the three sustainability dimensions. This has been done primarily through financing instruments committed to the environment and the generation of riches while taking present and future social welfare for the approval of each project. As stated on the official BBVA Mexico website (2022a, b, c), "For an entity to incorporate a green bond, the issuer, in addition to presenting the certification that guarantees the sustainability of the project, must share information about the use of resources and periodically report its activity and impact during the life of the bond."

There are different types of green bonds. According to Cabrera et al. (2021), the first among them is called Standard Green Use of Proceeds, which represents a benefit for the issuer having a backing in assets; the Green Revenue Bond, which translates into a debt obligation not secured by assets, that is, the credit exposure is to the committed cash flows whose use proceeds to related or unrelated green projects; the Green Project Bond, which is a bond for one or multiple green projects in which the investor has direct exposure toward the risk of that project, being insured by the project in which the resources, assets, and general balance are invested; and, lastly, the Green Securitized Bond, which is defined as a collateral bond to other specific green project(s), which can be covered bonds, asset-backed securities (ABS), mortgage-backed securities (MBS), among others, insured by green projects in general.

2.2 BBVA Mexico: Social Dimension

In accordance with the CSR's social dimension, it implies the shared and subsidiary commitment of investors, directors, collaborators, and suppliers for the care and promotion of workers' life quality, which promotes the full development of the communities in which they live (CEMEFI, 2022). In 2021, BBVA Mexico was awarded with the first place in the financial sector in the "Employers for Youth" survey, the Gallup award for "Exceptional Workplace"; positioned itself in the top 3 of the "Most Attractive Employers in Mexico" (list from Universum magazine), being a leader in the ranking of the 500 compànies against corruption; and the "Healthy and Responsible Company" distinction awarded by Workplace Wellness Council Mexico, among other awards that it has received as recognition of its social work.

Within the strategies of the financial group, it is established as a priority to form a diverse and empowered workforce, with a differential culture that internalizes the purpose and values of the group as well as provides opportunities for growth for all. Likewise, BBVA Mexico has also focused its efforts to contribute to the community to meet the Sustainable Development Goals (SDGs) of the United Nations, which is the reason the financial group increased sustainable financing by 373% in 2021 when compared to the year 2020.

In accordance with its commitment to achieve the United Nations' SDGs, the financial group, through its BBVA Mexico Foundation, works with four lines of action, these being social development, cultural promotion, strategic alliances, and operating expenses, in which during 2021 a total investment of 1,101,390,341 pesos was made. Some of the most important actions done with said investment were scholarships through the program "Becas para Chavos que Inspiran" ("Scholarships for Youth that Inspire," in English), donations to organizations that support education and art, donations of school supplies, rehabilitation of the social fabric in communities, promotion of art, culture, support of communities affected by natural disasters, and the "Together for Health" program in order to support the growing demand of hospital services. In the following, "Becas para Chavos que Inspiran" is described as one of the most successful programs.

2.2.1 The "Becas BBVA para Chavos que Inspiran" Program

Every year, the BBVA Mexico Foundation launches a national digital announcement to select the new scholarship-recipient generation, the "Chavos que Inspiran" program is directed to students starting their first year of junior high school in a public school, and with economic vulnerability, in addition to this, the scholarship program is constantly strengthened with allies from public and private universities, institutions specialized on disability, and 100% online education.

Fig. 1 Technical Junior High School No. 27 in Padilla, Tamaulipas. Source: Taken by the authors for this case study

The scholarship contains an economic component and integral formation support, according to the school level they currently attend, which is why the BBVA Mexico scholarship recipients have access to a digital platform in order to be informed of the online courses that are offered on different topics according to their level of studies; for example, the scholarship recipients in junior high school have a scholarship of $1,000.00 Mexican pesos per month and courses on financial education, digital abilities, and socioemotional skills, and students in high school level have a scholarship of $2,000 pesos in their first year, $2,500.00 in their second year, and $3,000.00 in their third year of studies and have access to courses on vocational guidance, study techniques, financial education, and socioemotional skills. Lastly, scholarship recipients in college have a monthly scholarship that goes from $3,500.00 Mexican pesos in their first year to $4,500.00 Mexican pesos in their last year of college and have training courses on financial education, leadership, English, and preparation for employment; they also count access to the mentorship program, in which, on a voluntary basis, a BBVA collaborator offers mentoring for students to assist on adapting to university and the path to the students' employment.

2.2.2 Testimony "Becas BBVA para Chavos que Inspiran"

Emilio Ochoa is a 14-year-old boy who is about to enter high school in a rural community called Nuevo Padilla, located in the state of Tamaulipas, in Mexico. He studied at the technical junior high school number 27, located in that community (Fig. 1). When interviewing Emilio, he was asked how he was able to obtain the BBVA scholarship, and he explained that on one occasion when he accompanied his mother to the said bank located in a nearby city, his mother realized that on televisions they were showing information about the scholarships for low-income

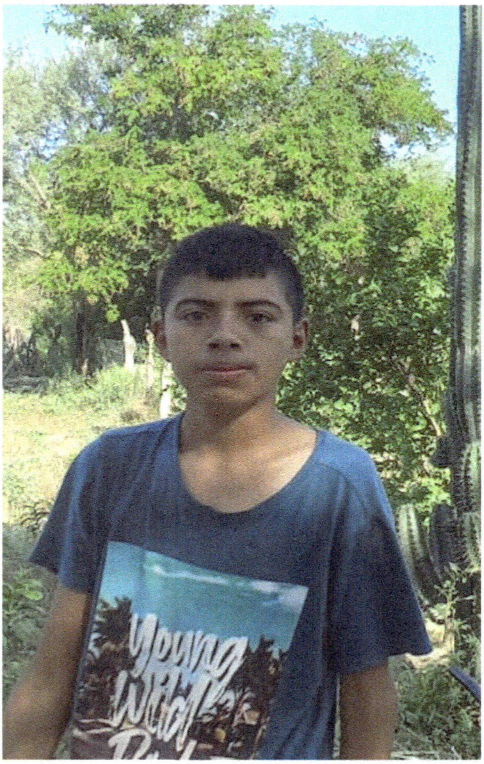

Fig. 2 Emilio Ochoa recipient of scholarship "Chavos que inspiran." Source: Taken by the authors for this case study

youth. Then, he noticed that his mother wrote the dates for the announcement, and she told him about the scholarship; however, he mentioned that his mother told him that it would be very difficult to get it, since she explained that there were many kids in that same situation, but they would try. After going back home, his mom went to a cybercafe to read the announcement from BBVA for the scholarships, and during the next two weeks, she used the bank's platform to send all the documents that were requested from Emilio's school. She registered her e-mail and constantly checked for a month; however, when she was disappointed because she was not receiving any notification, she received an email informing that her son Emilio had been selected for the scholarship.

Emilio Ochoa has been receiving the "Chavos que Inspiran" scholarship for three years, and now, as he is about to enter high school, he is very happy because they informed him that his scholarship would increase from $1,000.00 to $2,000.00 pesos per month, which will help him a lot, because to get from his house to the high school, he must travel approximately 10 kilometers every day, and that money will be useful to pay for the daily transportation as well as his food expenses (Fig. 2).

Inclusion: BBVA Mexico Financial Group, aware of the importance of inclusion and the access to education for all, has been strengthened with the training tools they call "Route to Success," managed by specialized advisors, who bring very close support to scholarship recipients on each school level coursed through the 100% digital platforms available to the said scholarship recipients.

To allow more students with disabilities to achieve their educational goals, BBVA Mexico Foundation through specialized institutions identifies and selects talented young people to support them with a scholarship and integral support. In 2021, 100 scholarships were distributed among junior high school, high school, university, and job training for scholarship recipients with some disability (BBVA Mexico, 2022a, b, c).

The success of the scholarship program is largely due to the alliances that BBVA Mexico Foundation has created with different organizations and universities, both public and private, including the National Autonomous University of Mexico (Universidad Nacional Autónoma de México), the National Polytechnic Institute (Instituto Politécnico Nacional), the National Autonomous University of Nuevo Leon (Universidad Autónoma de Nuevo León), the Technological University of Monterrey (Tecnológico de Monterrey), the Autonomous Technological Institute of Mexico (Instituto Tecnológico Autónomo de Mexico), and the University of Veracruz (Universidad Veracruzana), to name a few. This alliance scheme aims to establish collaborative relationships with universities, both public and private, foundations, institutions specializing in disability or digital formation, and with companies to bring better opportunities to talented young people.

To know the impact of the program, BBVA Mexico Foundation measures the results and effects generated on young people who are supported with these scholarships through impact evaluations done by external institutions.[1]

In addition to these impact studies, BBVA Mexico Foundation does and publishes annual results reports in which they inform through its website, social networks, and other official media, the results of the scholarship program in which they include number of scholarship recipients, the municipalities to which they belong, the donations received, and the alliances made, among other related topics. This is due to the Civil Association being self-committed to accountability, which is why it includes in its annual reports the financial statements audited by the firm KPMG (BBVA Mexico, 2021).

According to the 2021 annual report of the BBVA Mexico Foundation, since 2002, more than 353,000 scholarships have been handed out. Table 1 shows the number of scholarship recipients and mentors from 2017 to 2021. As can be seen, the number of scholarship recipients is increasing at all school levels, but in the years 2020 and 2021, the support at the university level was strengthened by increasing the number of scholarships from 7,504 to 11,421, equivalent to an increase of 52.2%. In

[1] Such is the case of Saint Martin and Szekely Pardo (2020), who analyzed the scholarship program's impact to know the probability that the scholarship recipients would have to graduate from the next level of education, for which they used the records of a standardized test taken by all students in Mexico to identify applicants who were still enrolled 3 years after starting the program, and their results show that BBVA Mexico's scholarship program had a positive and significative effect on junior high school completion.

Table 1 Number of scholarship recipients and mentors 2017–2021

School Level	2017	2018	2019	2020	2021
Jr High School	1,083	15,550	17,659	16,839	17,537
Highschool	3,061	11,973	13,535	16,342	15,245
University	637	1,737	4,273	7,504	11,421
Total	4,781	29,260	35,467	40,685	44,203
Mentors	1,478	2,222	2,326	5,055	7,824

Source: Own making with data found in the annual reports by Bancomer Foundation available on www.fundacionbbva.mx/pagina-inf/

addition, the report indicates that the people who graduated from the "Chavos que Inspiran" program had an average income 60.2% higher than people with similar profiles, and 58% of the female scholarship recipients who graduated managed to find their first job position with specialized profiles compared to the current employment of their parents. This was possible due to the support of 82 strategic allies and 7,824 voluntary mentors covering all the states of Mexico and over 1,520 municipalities. Likewise, the report informs that these actions contribute to the achievement of three of the SDGs (quality education, gender equality, and reduction of inequalities) (BBVA Mexico, 2022a, b, c).

2.3 BBVA Mexico: Environmental Dimension

Regarding the environmental dimension of the CSR, it implicates full responsibility for the environmental repercussions of the processes, products, and by-products that a company carries out, as well as specific actions to contribute to the preservation and improvement of the common ecologic heritage for the well-being of current and future humanity (CEMEFI, 2022). Regarding this dimension, the financial group has as a priority sustainability on a global level, which is why in 2021, it financed a total of 2,960 million euros to climate change.

In fact, BBVA Mexico's General Sustainability Policy defines and establishes the general principles and management goals in terms of sustainable development, within which it seeks the development of sustainable solutions, identifies opportunities, and offers counseling for customers and companies, for which it establishes two main actions focusing on terms of sustainability, these being climate change and inclusive and sustainable social development. In relation to climate change, it focuses on three business initiative categories, these being energy efficiency, circular economy, and the reduction of carbon dioxide (CO_2).

Some of the actions carried out by BBVA Mexico are focused on the consumer, and to contribute to sustainable development, they promote the use of hybrid and electric cars by offering fixed and preferential rates to customers who require credit to purchase this type of cars. In the same way, it provides extra benefits when hiring insurance for hybrid or electric cars. In addition to these actions that contribute to caring for the environment, they also offer a sustainable investment fund known as

"BBVAESG", which promotes investment on companies with a high environmental, social, and corporate governance rating, avoiding exposure to industries involved on environmental issues, thus looking to reduce carbon emissions.

Regarding the actions focused on the business sector, the financial group grants sustainable import letters of credit to those companies that desire to acquire merchandise, machinery, and equipment called "green" such as solar panels, electric equipment and motors, wind generators, and LED lights, among others. Another program with which BBVA Mexico has supported entrepreneurs is called "Demandable Guarantees and StandBy," which supports businesspeople to participate in public international licitations for the adjudication of sustainable projects focused on water treatment, energy efficiency, renewable energy, and the conservation of biodiversity. This is how the financial group fulfills its commitment to the United Nations' SDGs and is committed with Mexico's sustainable development with concrete actions.

3 Conclusions

The ecological, economic, and social crisis that has been worsening over the years worldwide has been due to the way that businesses are carried out in a "business as usual" way, which is generating a paradigm shift globally. This new paradigm has been called Sustainable Development, in which it is recognized that economic activities, social development, and environmental care are equally important for a society. The foregoing has stimulated various countries and international organizations to form agreement points such as the 17 Sustainable Development Goals, in which it is sought in a certain way to create a social and environmental awareness about the way that things are produced and consumed. In this way, it seeks to generate business models that are compatible with sustainable development.

From the business management approach, the study of responsible behavior of the company has been widely approached, especially in matters that have to do with the creation of programs that promote the care for resources, workforce well-being, as well as the impact that its activities generate on the environment, among others. In this case, evidence has been presented on how one of the most important financial companies in the country has implemented diverse strategies, actions, and programs that have led it to achieve national and international certifications on different categories to classify itself as a socially responsible company. The above actions have been carried out despite that there is no conclusive evidence on the benefits that the financial group might obtain by complying with this type of social activities that it carries out voluntarily.

The BBVA Mexico financial group has gone beyond complying with its legal obligations and has decided to contribute to its community in different ways, which include issuing green bonds, scholarships for low-income students, support in natural disaster situations, the rehabilitation of public and private spaces in highly marginalized urban areas, the implementation of programs that promote culture and art, and actions to support hospitals through the "Together for Health" program, among others. However, it is also true that if they were to fail with complying with this idea of being a socially responsible company, they may run the risk of losing a good share of the market.

4 Discussion Questions

Question 1. What is the importance of financial groups developing socially responsible actions?

Question 2. Regarding the multiple sectors and their business activity, what are those that can enhance their positive impact with the support of financial groups such as BBVA Mexico? Why?

Question 3. Do you think there are conflicts of business interest to develop socially responsible actions in the financial sector? Why? Could you comment on an example case?

Question 4. The adherence to the seal of a socially responsible company is carried out voluntarily; do you consider that it should be mandatory in general? Why?

Question 5. Which other productive sectors should create organizations for the cooperation regarding the goals of the 2030 agenda like the financial sector did with the UNEP FI? What type of joint strategy would be the most convenient?

Question 6. Lastly, how can responsible consumption contribute to companies that develop high-impact sustainable projects?

Appendix: Teaching Notes for the Case

Synopsis

This case study exposes the actions developed by BBVA Mexico Financial Group, which by its actions encourage a sustainable business and social model. BBVA Mexico increased sustainable financing by 373% in 2021, causing the emission of five green bonds and two social bonds by the International Capital Market Association, in accordance with the framework for the emission of bonds linked to the Sustainable Development Goals (SDG). BBVA Mexico counts with programs for enhancing gender equality, optimal work conditions, scholarships for Mexican youth, the promotion of renewable energy sources, and the fixation of intermediate objectives for the decarbonization of the investment wallet in four different industries which are intensive regarding the emissions of carbon dioxide (CO_2); BBVA Mexico Financial Group obtained the first place on the ranking of the Most Responsible Companies in Mexico within the financial sector and overall the third place on the "Merco Responsibility ESG (Environmental, Social and Governance)," consolidating themselves as one of the leading banks regarding the scope of sustainability.

Potential Audience and Instructor's Material

This case study has been designed for analysis purposes from the perspective of marketing and corporate social responsibility, in alignment with the Sustainable Development Goals' agenda. As main audience for this chapter, the following are suggested:

- Students of professional careers linked to marketing and subjects with contents of corporate social responsibility
- Postgraduate students, whose contents have an emphasis on corporate social responsibility and strategic business marketing
- Public in general with interests for reflecting over related topics

Learning Objectives

1. To understand the convergence of corporate social responsibility and sustainability
2. To understand the sustainable impact that promotes responsible actions through third parties involved
3. To analyze the importance of responsible consumption through a company with a high impact in the financial sector
4. To identify the main socially responsible actions of a financial group that leads in its area
5. To ponder on the importance of establishing strategic alliances with the implicated groups and to generate a real change with a higher impact in terms of social responsibility for a sustainable lifestyle

Time Frame for Class Discussion

This case analysis is planned for 60–100 min, depending on anticipated reading work and the number of students within the class group.

Suggested Discussion Questions

1. Question 1. What is the importance of high reach financial groups developing socially responsible actions?
2. Question 2. Regarding the multiple sectors and their business activity, what are those that can enhance their positive impact with the support of financial groups such as BBVA Mexico? Why?
3. Question 3. Do you think there are conflicts of business interest to develop socially responsible actions in the financial sector? Why? Could you comment on an example case?

4. Question 4. The adherence to the seal of a socially responsible company is carried out voluntarily; do you consider that it should be mandatory in general? Why?
5. Question 5. Which other productive sectors should create organizations for the cooperation regarding the goals of the 2030 agenda like the financial sector did with the UNEP FI? What type of joint strategy would be the most convenient?
6. Question 6. Lastly, how can responsible consumption contribute to companies that develop high-impact sustainable projects?

Analysis

Answer to Question 1. *What is the importance of high reach financial groups developing socially responsible actions?*

- The financial sector has a high socioeconomic impact just for the fact that they offer services and resources for the management of money. In addition to this, those companies that are distinguished as socially responsible promote economic growth from their efforts and policies, but they also build a paradigm shift so that the resources that are linked to national and international transactions are directed toward sustainability where resources of future generations are better distributed.

Answer to Question 2. *Regarding the multiple sectors and their business activity, what are those that can enhance their positive impact with the support of financial groups such as BBVA Mexico? Why?*

- Any sector that seeks financial services in general, suggesting a high growth potential for economic entities that project themselves toward innovation in order to generate a smaller environmental footprint or the solution to environmental and social issues that afflict us.

Answer to Question 3. *Do you think there are conflicts of business interest to develop socially responsible actions in the financial sector? Why? Could you comment on an example case?*

- Conflicts of interest are one of the main detractors in this kind of policies in favor of sustainability. The paradigm of traditional interests must be broken, and when this happens, sustainable interests will be a priority; legitimizing the concept of social responsibility, for example, through consumption, which is what ultimately guides global supply chains.

Answer to Question 4. *The adherence to the seal of a socially responsible company is carried out voluntarily; do you consider that it should be mandatory in general? Why?*

- This is a discretionary question. Before answering, it is suggested to analyze the current context in which we live so as to be able to decide whether voluntary or obligatory actions are required in terms of social responsibility and responsible consumption.

Answer to Question 5. *Which other productive sectors should create organizations for the cooperation regarding the goals of the 2030 agenda like the financial sector did with the UNEP FI? What type of joint strategy would be the most convenient?*

- It is suggested that all sectors and governments, together with their stakeholders or interest groups, should converge on a single sustainable paradigm for this to be possible. This way, the programs that impact the different dimensions of sustainability could be developed more articulately within the current national and international environment.

Answer to Question 6. *Lastly, how can responsible consumption contribute to companies that develop high-impact sustainable projects?*

- Anyone interested in obtaining a credit or financial services in general should be informed about the policies and programs that the providers in the market are carrying out. This way, the socially responsible cycle would be supported through consumption, with whether you are a natural or a legal person not mattering.

Final thoughts: The issue of corporate social responsibility can be understood from various dimensions; however, this construct converges with sustainability from the economic, social, and environmental perspectives. Although the social and environmental dimensions can be translated into a philanthropic base, the economic dimension is the one that maintains the current discussion regarding how is it possible for companies to meet their social and environmental responsibility without neglecting the profitable responsibility toward the main groups involved?

In this sense, the CSR of companies tries to align with the demands of such groups from sustainable models that can maintain profitability through beneficial strategies from most environmental angles and for most related social strata. The BBVA Mexico case is an example of the contribution that can be managed from the financial sector, in conjunction with international plans and programs that are aimed toward a sustainable development. This is not a minor thing if one considers that this group has not implemented said strategies in an isolated way but that they have formed alliances with third parties involved that are interested in aligning themselves to the sustainable cause and meeting the demands of the global agenda to enable the continuity of their business activities.

The dimension that supports the business dimensional pyramid corresponds to the economic dimension. So, how can each segment and profile involved in society contribute so that these types of sustainable efforts have a greater positive impact on the three dimensions? Based on this case study, students are encouraged to carry out research on actions that can be implemented from any role in society, in relation to green projects that remain current in the context or country in which they are, thus contributing to greater support so that projects with green foundations can be carried out successfully.

References

BBVA Mexico. (2021). *Informe anual 2021*. Retrieved from https://www.bbva.mx/content/dam/public-web/mexico/documents/negocio-responsables/BBVA-IA2021_Final.pdf

BBVA Mexico. (2022a). *Becas BBVA para chavos que inspiran*. Retrieved on March 5th, 2022, from https://www.fundacionbbva.mx/becas-bbva-para-chavos-que-inspiran/

BBVA Mexico. (2022b). *Estrategia*. Retrieved on March 25, 2022, from https://portal.bbva.mx/informes/2021/estrategia

BBVA Mexico. (2022c). *Resultados convocatoria 2021*. Retrieved on March 5th, 2022, from https://storage.googleapis.com/bbva-mx-becas-fundacion.appspot.com/PDFINCLUSION.pdf

Cabrera, L. F., Guzmán, C. S. E., Zetina, Q. A., & Castillo, F. J. (2021). Bonos verdes. Creando sostenibilidad en las finanzas. Revista Divulgación. *Economica, 1*(1).

CEMEFI. (2022). *El concepto de responsabilidad social empresarial (s.f.)* Retrieved on March 21 2022 from https://www.cemefi.org/esr/images/stories/pdf/esr/concepto_esr.pdf

Sarro Álvarez, M.M., Cuesta Valiño, P., & Penelas Leguía, A. (2007). La responsabilidad social corporativa (RSC), una orientación emergente en la gestión de las entidades bancarias españolas. *Conocimiento, Innovación y Emprendedores*: Camino Al Futuro, 1928–1942. Retrieved from https://dialnet.unirioja.es/servlet/articulo?codigo=2232692

García, M. E. R., & Márquez, S. M. (2021). Estimating the Medium-Term impact of a Scholarship program using. *Brazilian Journal of Business, 3*(1), 49–77. https://doi.org/10.34140/BJBV3N1-004

INEGI. (2020). *INEGI presenta resultados de la encuesta para la medición del impacto covid-19 en la educación (ecovid-ed) 2020 [Comunicado de Prensa]*. Retrieved from https://www.inegi.org.mx/contenidos/saladeprensa/boletines/2021/OtrTemEcon/ECOVID-ED_2021_03.pdf

NOTICIAS ONU. (2019). *Los desastres son la nueva normalidad*. Retrieved from https://news.un.org/es/interview/2019/11/1465021#:~:text=En%20la%20actualidad%2C%20el%2090,efectos%20devastadores%20de%20las%20cat%C3%A1strofes

Saint Martin, M., & Szekely-Pardo, M. (2020). Corporate social responsibility (CSR) in the banking sector. *Journal of Education and Learning, 9*(1), 172–195.

SENER. (2018). *Balance general de energía 2018*. Retrieved from https://www.gob.mx/cms/uploads/attachment/file/528054/Balance_Nacional_de_Energ_a_2018.pdf

UNEP FI. (2022). *Acerca de*. Retrieved from https://www.unepfi.org/about/

Sustainable Cosmetics and Perceived Brand Image on Social Networks: Comparison Between Millennial and Centennial Women

César Sahelices-Pinto, Ana Lanero Carrizo, and Jose Luis Vazquez Burguete

Learning Objectives

1. To compare the campaigns developed in social networks to create a brand image linked to sustainability both by companies specifically offering sustainable products and by those 'conventional' ones (i.e. not offering sustainable products or not only) in the cosmetics sector
2. To study the positioning generated among consumers by 'sustainable' cosmetic brands from their campaigns in social networks
3. To reflect on the usefulness of the campaigns occasionally developed by conventional 'cosmetic' brands in social networks to decouple their positioning from other dimensions of beauty and health
4. To compare the positioning generated by the sustainable-image creation campaigns in social networks conducted by 'sustainable' and 'conventional' brands among consumers of different generational profiles (millennials or centennials)

1 Introduction

The cosmetics sector is one of the areas in which there has been a greater increase in the consumption of products from sustainable brands in recent years (Fortuny et al., 2020). These products are characterised by having a composition based on ingredients that are respectful with the environment and people's skin, since in their preparation, they avoid the use of compounds derived from petroleum and other chemicals and silicones. Conventional products are not only obtained from such harmful substances, but they are also difficult to recycle and use to end up in waters and oceans, thus seriously polluting the ecosystem and the marine

C. Sahelices-Pinto (✉) · A. Lanero Carrizo · J. L. Vazquez Burguete
Faculty of Economics and Business Sciences, University of León, León, Spain
e-mail: cesar.sahelices@unileon.es; ana.lanero@unileon.es; jose-luis.vazquez@unileon.es

E. C. Bianchi et al. (eds.), *Responsible Consumption and Sustainability*, Springer Business Cases, https://doi.org/10.1007/978-3-031-30742-3_14

environment. All the experimentation animals that suffer and die when used to speed up the process of launching cosmetic products should also be taken in consideration to accurately assess the situation (Cosmetics Europe, 2017).

Nowadays, sustainable cosmetics represent a small share of the Spanish market if compared to conventional cosmetics. However, the level of consumption of cosmetic products from sustainable brands is growing as in 2017–2018 sales of sustainable cosmetic products increased by 20%. Moreover, in 2018, they registered a turnover of more than 800 million euros, that is, 12% of the total beauty products in the Spanish market (Fortuny et al., 2020). In this sense, it is estimated that 30% of cosmetic market launches are linked to the natural or sustainable product category. Nevertheless, the relevance of sustainable cosmetics in Spain is still below that of countries like Germany or France (Stanpa, 2020).

This increasing demand for sustainable cosmetic products puts pressure on companies to be more ethical in social and environmental terms, according to the spreading of a concept of 'clean beauty'. So, there has also been a proliferation in recent years of communication campaigns aimed at making brands' commitments to sustainability visible, so that they are easily perceived as such by consumers. Specifically, social networks are currently one of the main communication tools at the service of cosmetic companies to connect their brands with their target audience.

In this regard, it should be noted that not all consumers value the sustainability of the products they purchase to the same extent, nor they are exposed in the same way to the communication actions of companies through their social media channels. Thus, it is known that the main determinant for both issues has to do with the existence of different generational cohorts (Eixarch et al., 2019; Swanepoel et al., 2009), which is often considered by companies when making decisions about market segmentation.

The search for a brand image linked to sustainability through campaigns in social networks is a common practice for cosmetics firms committed from the beginning to these principles (e.g. Lush Fresh Handmade Cosmetics and Freshly Cosmetics) and for conventional brands that seem to look for a change in a positioning which was traditionally more linked to 'only' the values of health and beauty (such is the case of *L'Oréal*, *Maybelline*, *Dove* or *Clinique*). However, are social media campaigns just as effective in conveying a sustainable brand image regardless of the brand's previous trajectory? To what extent is this communication perceived as image laundering and generates scepticism in the public? And, also related to both issues, are there differences in the perception of brands by consumers of different generations?

To contribute to this debate, this case study analyses the positioning of different sustainable and conventional cosmetic brands according to the perception of two different generations of young consumers (millennials and centennials), based on the brand image built through online advertising in social networks.

2 Case Development

2.1 Brand Image and Social Networks

Differentiating goods or services based on their visible characteristics becomes more and more difficult day by day. Consequently, the importance of brand image increases to guarantee the success and competitiveness of a company in the market. The brand image appears as 'the perception of a brand as a reflection of the associations that a consumer makes of it in their memory' (Keller, 1993, p. 3). Different types of brand associations may be present in the mind of a consumer, specifically in terms of the following: (i) attributes related and non-related to products; (ii) their functional, experiential, and symbolic benefits; and (iii) attitudes about the general assessments of consumers about a concrete brand. Such associations arise from the consumer's general exposure to marketing activities and brand interactions (Keller, 2003).

Based on the previous definition, it can be stated that having a sustainable brand image appears in the mind of consumers as a synonym of being associated to economic, social, and environmental commitments and concerns. In other words, launching sustainable products is not enough to develop a sustainable brand image and reputation, but they should be perceived as such by consumers. In commercial practice, the promotion of a sustainable brand image can be done by using various green marketing strategies and tools. Specifically, eco-labels and advertising campaigns are a worthwhile option, which in recent years have been strongly characterised by the rapid growth experienced by social networks and their high use by the younger generations.

The cosmetics industry has benefited from social media platforms by connecting brands with consumers and enabling social integration and transparency about goods and services. Maybe platforms like Instagram and YouTube do not influence all age groups, but they do contribute to generate demand for beauty products. Evidence shows that more than 50% of beauty videos posted on YouTube are related to cosmetics (Statista, 2022), as it is clear that social networks benefit the cosmetics industry by spreading electronic word-of-mouth (eWOM), which helps to promote a company's product and its brand/s awareness (Shen & Bissell, 2013).

Furthermore, eWOM in social networks usually results in viral marketing, and companies can take advantage of rapid and massive communication of knowledge. People engagement in Facebook amplifies the way in which businesses interact with customers by enabling them to respond to market changes and demands. In addition, eWOM application has become an important part of the marketing strategies of corporations to communicate a brand image or values, as well as to increase customer loyalty. Similarly, eWOM seems to have a significant effect on consumers' purchase intention and attitude toward a brand. Therefore, businesses can use social media platforms to post and share content that builds support for their brands.

Image 1 Logos of *L'Oréal*, *Maybelline*, *Dove*, *Clinique*, *Lush* and *Freshly Cosmetics*. Source: L'Oréal (2022), Maybelline (2022a), Dove (2022a), Clinique (2022), Lush (2022a), and Freshly Cosmetics (2022a)

2.2 Campaigns for Sustainable and Conventional Cosmetic Brands

In the current cosmetics market, brands that have emerged with a clear commitment to sustainability coexist with more 'conventional' brands which only occasionally include sustainability issues in their regular commercial campaigns. In this sense, we can refer to six cosmetic brands appearing as particularly relevant in customer perception:

1. *L'Oréal* is the world's largest cosmetics company, and it was established in France at the beginning of the twentieth century. Thanks to an innovative nature, the company has incorporated digital technology into its products long before its competitors, which has contributed—and still is—to a quick adaptation to the constantly changing lifestyles and needs of consumers (Image 1).
2. *Maybelline* was established at the beginning of the twentieth century as a small family business focused on mascara. Nowadays, it is a worldwide recognised cosmetic brand due to its New York-inspired products with a remarkable value for money.
3. *Dove* is the line of personal care products belonging to the multinational Unilever, whose first product was launched in the United States at the mid-1950s. The brand is easily recognisable for its mission, aiming to break with prevailing aesthetic stereotypes and broaden the definition of beauty.
4. *Clinique* is an American brand that emerged in the late 1960s. It was initially developed by dermatologists, and its current products maintain the commitment to not include allergens, irritant components, or ingredients that could damage the skin.
5. *Lush* was established at the end of the twentieth century in the United Kingdom as a company devoted to natural, sustainable, and vegan cosmetics. It stands out because its production is handmade to a great extent as well as due to the fact that

Image 2 Post of *Clinique* on Instagram. Source: Clinique (2021)

some of its references require to be kept refrigerated for its correct conservation as they incorporate fresh ingredients.

6. *Freshly Cosmetics* is a Spanish natural cosmetics firm established at the second decade of the twenty-first century by a group of chemical engineers. The products of the company contain only healthy, sustainable, and vegan ingredients.

Going into detail in terms of the image that brands reflect on social networks, it can be seen, for instance, that *Clinique* carries out inclusive advertising by incorporating women with skin imperfections (Image 2).

As a result of a different approach, it could be said that *Maybelline* appears as a conventional cosmetic brand that tests on animals and currently also includes images of women with imperfections in its advertising campaigns (Image 3).

L'Oréal is another brand with a long record of communicating attributes of perfection in women (Image 4). However, since 2020, the brand has placed

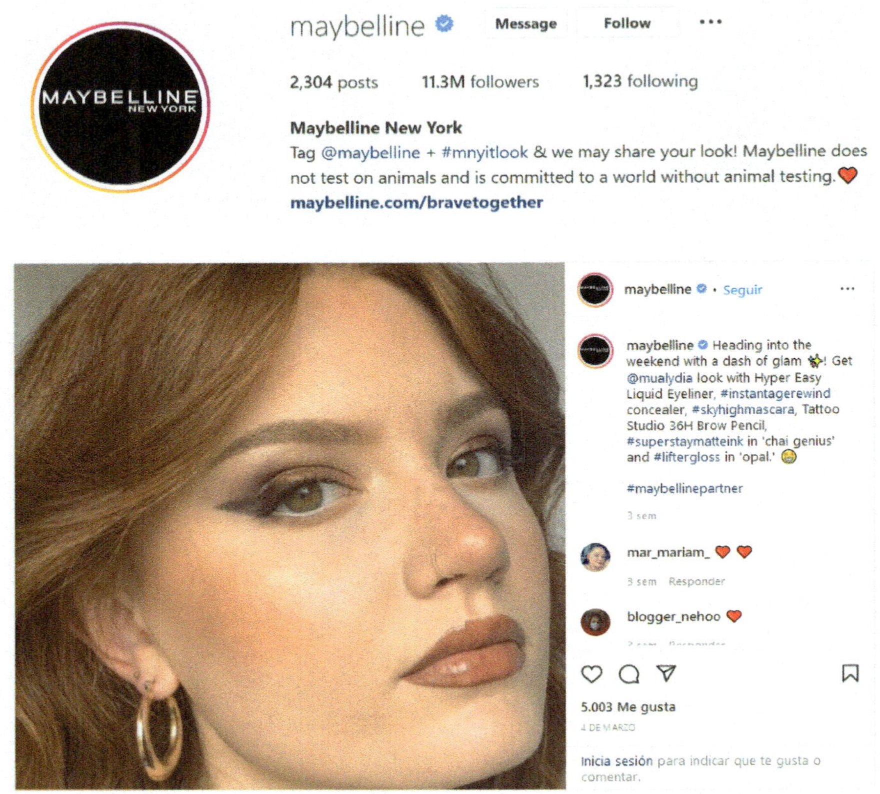

Image 3 Post of *Maybelline* on Instagram. Source: Maybelline (2022b)

sustainability at the core centre of its business and communication strategies by carrying out social and environmental actions.

Dove is shown as a sustainable cosmetic brand because it does not test on animals and it uses recycled plastic bottles in its packaging and has social initiatives such as the *Dove Day*, aiming to help the next generation of individuals—both men and women—to develop a positive image of themselves, thus empowering both genders (Image 5).

As for *Lush* (Image 6) and *Freshly Cosmetics* (Image 7) brands, they both have already been built since their inception on the principle of sustainability. It should also be noted that *Lush* does not develop its communication on social platforms to a great extent.

As we can see, this group includes some brands that have been sustainable since their inception and whose communication has reflected inclusiveness and transparency from the very beginning (*Freshly Cosmetics* and *Lush*), with other brands that initially did not use sustainable attributes, did not claim diversity of bodies, or did

Image 4 Post of *L'Oréal* on Instagram (*'We create the beauty that moves the world'*). Source: L'Oréal (2021)

not seek positive impacts on society in their campaigns, but they do now (*L'Oréal*, *Dove* and *Clinique*). There is also a brand that even today is not appearing as socially conscious, as it does not give relevance enough to those principles in the definition of its organisational mission (*Maybelline*).

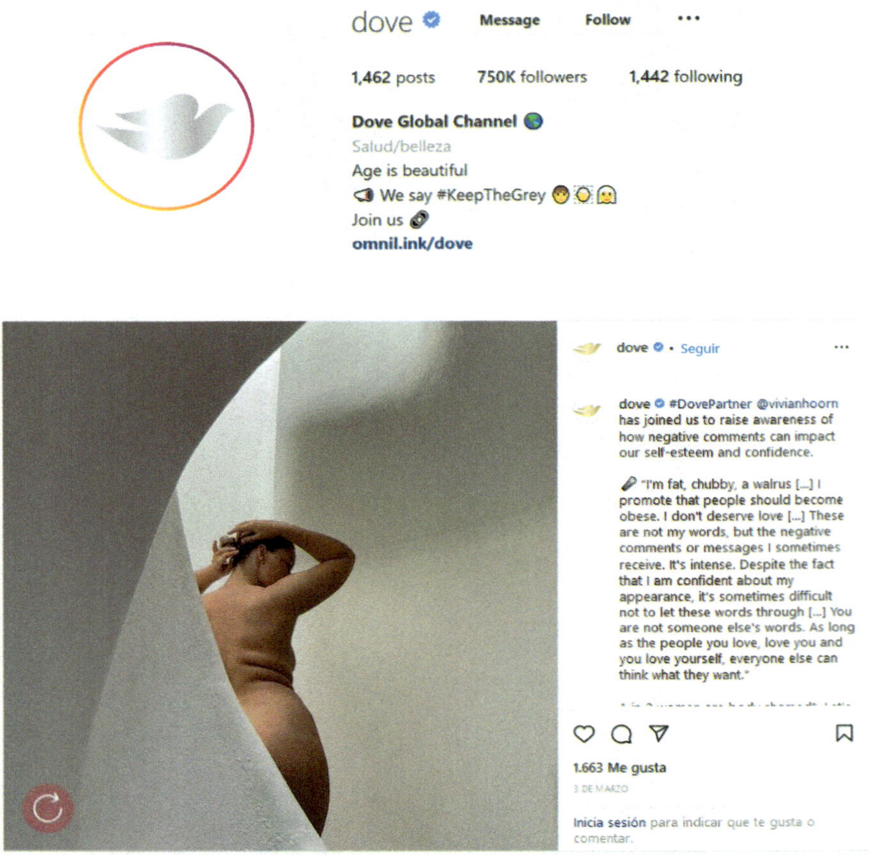

Image 5 Post of *Dove* on Instagram. Source: Dove (2022b)

Focusing on social networks, six of the most visible attributes at the moment are *perfection*, *femininity*, *real women*, *sustainability*, *absence of stereotypes* and *empowerment*. By analysing these attributes and their relationship with the brands, it is possible to know, on the one hand, if the brands know how to effectively communicate in their advertising the characteristics that make them sustainable and, on the other hand, what is the perception of consumers about these either 'sustainable' or 'conventional' cosmetic brands.

Image 6 Post of *Lush* on Instagram Source: Lush (2022b). (*'In a better place'*)

2.3 Positioning of Cosmetic Brands for Millennials and Centennials

Once we have exposed the efforts of different sustainable and traditional cosmetic brands to link their image to sustainability attributes through their publications on social networks, it is worth analysing to what extent they are able to translate them into the expected positioning according to the perception of their target audiences. At this purpose, it should be noted that consumers do not make up a homogeneous audience, but different groups of consumers are likely to retain different mental approaches to sustainability. For example, people belonging to different generational cohorts often attribute different meanings to advertising campaigns, due to their

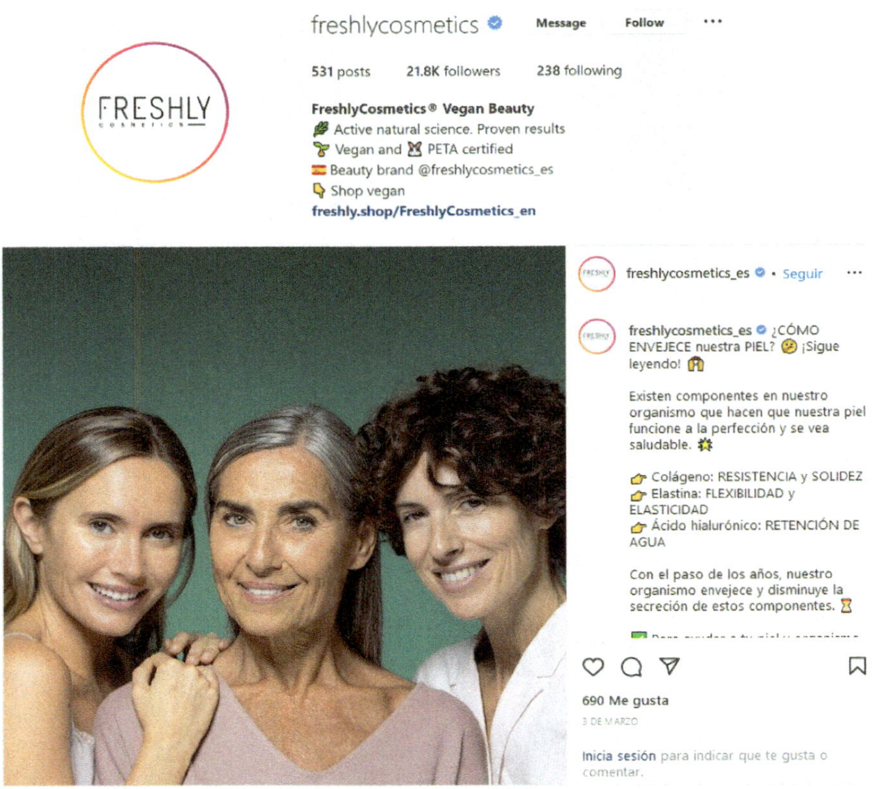

Image 7 Post of *Freshly* Cosmetics on Instagram. Source: Freshly Cosmetics (2022b)

changing needs and buying habits. Definitely, shopping conveys different meanings and forms for different generations.

Similarly, the relationship between each generation and the media also influences the promotional strategies and their implementation. Just due to it, while for the youngest generations, i.e. millennials (or Generation Y, including those individuals born between 1981 and 1994) and centennials (or Generation Z, those born between 1995 and 2010), the Internet is now an important issue in daily practices and routines, this is not necessarily the situation for other generations (Swanepoel et al., 2009). In any case, the truth is that the Internet and mobile advertising have become unavoidable promotional tools for most brands.

However, although both millennials and centennials represent two cohorts of young consumers immersed in the digital world, it seems that there are differences between them when it comes to sustainability issues. Namely, and in a general sense, it has been suggested that centennials are more environmentally conscious than millennials (Eixarch et al., 2019). In fact, the members of the Generation Z (or centennials) are increasingly known as the 'green generation', as they demonstrate environmentally and socially conscious consumer behaviour and translate their awareness into eco-friendly and organic product choices more than millennials do. This is due to the increase in their purchasing power and their digitally enhanced access to information. Specifically, market observers believe that the linkage of consumer acts with ethical and ecological choices is becoming the new status symbol for centennials and an expression of their identity, rather than in the case of millennials (Composed Creative, 2019).

Applying these postulates to the analysis of the positioning of sustainable and conventional cosmetic brands presented in the previous section, it might be expected that even though millennials and centennials are equally exposed to advertising campaigns on social networks, these could be perceived differently, so that the youngest consumers could be—or should be—more sensitive to the image of sustainability that is sought to be built. In order to contrast this idea, a small survey was carried out by contacting a group of women through social networks.[1]

The comparative analysis between millennials and centennials is shown in Figs. 1 and 2, [2] where the different cosmetic brands are located along with the various valued attributes. As it can be observed, the perceived brand image built based on the indicated attributes does not differ to a great extent between the study participants, regardless of the generation the individuals belong to.

So, on the one hand, the conventional cosmetic brands *Maybelline*, *L'Oréal,* and *Clinique* are perceived as remarkably similar to each other in that both groups are close to the attributes of 'perfection' and 'femineity'. These brands are also perceived as being distant from the others in terms of attributes as 'sustainability', 'empowerment', 'without stereotypes' and 'real women'.

On the other hand, the sustainable cosmetic brands *Lush* and *Freshly Cosmetics* are also perceived as remarkably similar to each other by the participants regarding the characteristics of 'sustainability' and 'empowerment'. However, these brands are not identified with the rest of the considered attributes, moving away from competing brands (*Maybelline*, *L'Oréal*, *Clinique* and *Dove*).

Finally, the brand *Dove* is perceptually located close to the attributes of 'real women' and 'without stereotypes', as well as quite distant from the rest of the characteristics and brands, which gives it a certain particularity.

---

[1] The sample consisted of 237 women, of which 135 and 102 declared to be millennials (between 28 and 40 years old) and centennials (between 18 and 27 years old), respectively. From the total figure, 45% indicated that they were employees, 41% were students or interns, and 8% were self-employed.

[2] The obtained data were analysed with the statistical package IBM SPSS Statistics version 26. The multivariate technique selected to address the research objective was simple correspondence factor analysis, which was performed twice, one for the subsample of millennials and another for the centennials.

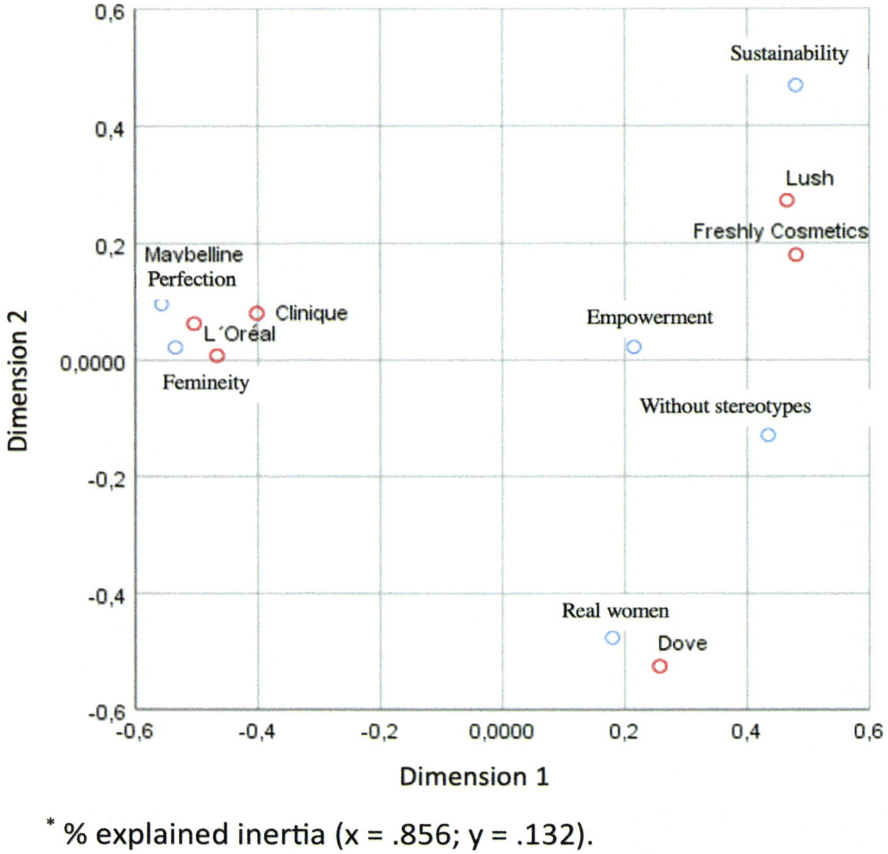

Fig. 1 Perceptual map of brands and attributes for millennials*. *% explained inertia ($x = 0.856$; $y = 0.132$). Source: own elaboration

3 Conclusions

As it has been verified, the perceptual maps of the image of the different cosmetic brands have revealed that both millennials and centennials have very similar mental representations of them. Therefore, although centennials are known to have a greater sustainability awareness than millennials (Eixarch et al., 2019), it seems that the marketing actions performed by cosmetic brands in social networks have a similar repercussion on the new generations of buyers.

Specifically, and according to data, conventional cosmetic brands *L'Oréal*, *Maybelline* and *Clinique* are associated with the advertising attributes of 'perfection' and 'femininity' and distant from the ideas of 'real women' or 'sustainability'. *Clinique* is considered as a conventional cosmetic brand, and it is perceived as being distant

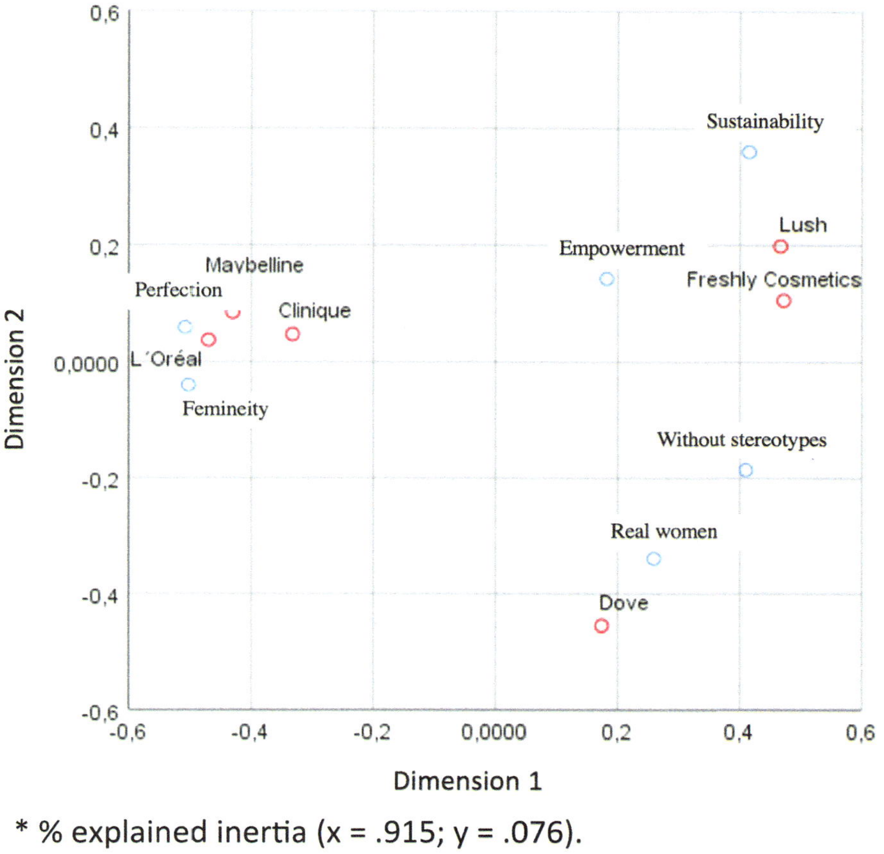

Fig. 2 Perceptual map of brands and attributes for centennials*. *% explained inertia ($x = 0.915$; $y = 0.076$). Source: Own elaboration

from the values of sustainable cosmetics. Similarly, despite that *Maybelline* has oriented its marketing strategy to get it closer to a young target public, the real fact is that users continue to associate its image with attributes unrelated to sustainability. Even *L'Oréal* and *Dove* continue to be considered brands that are far from their current sustainability values, which points to their communication strategies to transmit those values are not being effective.

Regarding *Lush* and *Freshly Cosmetics*, they are efficiently associated with sustainability attributes in their advertising campaigns (although this evidence could be interpreted as indifference when making a decision to choose between the one or the other brand). This is especially interesting in the case of *Lush*, since it has been able to successfully communicate its sustainable attributes and it has succeeded in being perceived as such without turning to social networks.

Even when sustainable cosmetics achieve more followers day by day, the truth is that it continues to be a sector linked to a segment of minimal consumption when compared to conventional cosmetics. In any case, it should not be forgotten that competition between cosmetic brands is fierce and that in view of advertising and communication trends, sustainability, environmental awareness and social concern appear as key points to distinguish products, attract the audience and retain it.

Regardless of this, this case study indicates that consumers persistently perceive the brand image. So, conducting punctual communication campaigns supported by sustainability issues is not probably enough to achieve a decisive change in perception, an issue that could be generalised to any young target audience. Moreover, such campaigns could even have adverse consequences by generating scepticism towards the brand.

In short, commitment to social and environmental values in communication and social platforms is a strategy that nowadays allows building a solid brand image in the short and medium terms. However, if this strategy is not supported in real business management actions, there is a risk that it could be interpreted as greenwashing.

4 Discussion Questions

1. In view of the conclusions of the case, what are the keys to building a sustainable cosmetic brand positioning on social networks?
2. In the case of brands with an image traditionally linked to other attributes, how can they reach consumers who are sensitive to sustainability? How to give value to the commitments assumed with society and the environment, even if they are below those of other sustainable cosmetic brands?
3. Is it necessary to work on empowering consumers against greenwashing campaigns or do they have their own resources to judge them appropriately?
4. To what extent can a segmentation between millennial and centennial consumers be useful in the sustainable cosmetics sector? And in other contexts?
5. What are the limitations of the positioning study presented in the case, and how could the situation be improved?

Appendix: Teaching Notes for the Case

Synopsis

Nowadays, cosmetics represent one of the sectors in which the growing demand for sustainable product lines is most appreciated. Due to this reason, different companies seek to link sustainability to their brands through social media campaigns, a practice common both to cosmetics firms linked from their very beginning to such sustainability principles and to more conventional brands, traditionally more linked to health and beauty values, and looking for a change in their positioning.

Regarding the latter, it is worth asking whether these campaigns are equally effective in transmitting a sustainable brand image despite the brand's previous trajectory or, on the contrary, they are perceived as image laundering, then generating rather scepticism in the market. Doubts also arise on the similarity (or not) of the perception of brands by consumers, regardless of the generation they belong to.

In this context, the positioning of several sustainable and conventional cosmetic brands is analysed, considering the generation of belonging of a group of young female consumers (millennials and centennials) and their perception of brand image built through advertising on social networks.

Potential Audience and Instructor's Material

The case has been developed for use in general courses on marketing and consumer behaviour as well as in specific seminars on corporate social responsibility and sustainable consumer behaviour.

The potential audience for this case includes both undergraduate students pursuing degrees in marketing or business administration and graduate students pursuing MBA and Executive Programs about CSR and sustainability.

Learning Objectives

1. To compare the campaigns developed in social networks to create a brand image linked to sustainability both by companies specifically offering sustainable products and by those 'conventional' ones (i.e. not offering sustainable products or not only), in the cosmetics sector
2. To study the positioning generated among consumers by 'sustainable' cosmetic brands from their campaigns in social networks
3. To reflect on the usefulness of the campaigns occasionally developed by conventional 'cosmetic' brands in social networks to decouple their positioning from other dimensions of beauty and health
4. To compare the positioning generated by the sustainable-image creation campaigns in social networks conducted by 'sustainable' and 'conventional' brands among consumers of different generational profiles (millennials or centennials)

Time Frame for Class Discussion

This case can be taught in a session lasting 60–90 min. The concrete duration depends on the number of students participating in the discussion activities as well as on the previous work that they have done before the session.

Suggested Discussion Questions

1. In view of the conclusions of the case, what are the keys to building a sustainable cosmetic brand positioning on social networks?
2. In the case of brands with an image traditionally linked to other attributes, how can they reach consumers who are sensitive to sustainability? How to give value to the commitments assumed with society and the environment, even if they are below those of other sustainable cosmetic brands?
3. Is it necessary to work on empowering consumers against greenwashing campaigns, or do they have their own resources to judge them appropriately?
4. To what extent can a segmentation between millennial and centennial consumers be useful in the sustainable cosmetics sector? And in other contexts?
5. What are the limitations of the positioning study presented in the case and how could the situation be improved?

Analysis

Answer to Question 1. *In view of the conclusions of the case, what are the keys to building a sustainable cosmetic brand positioning on social networks?*

Students can work individually or in small groups to answer this question. Then, they can discuss their viewpoints with the whole class, under the guidance of the teacher.

Some issues to be taken into account when analysing the effectiveness of sustainability campaigns in social networks are as follows:

- Factors related to the company: mission, values and objectives, sustainability commitment, brands and starting image, products and starting image, etc.
- Factors related to the target audience: personal variables, situational variables, sensitivity towards sustainability, purchasing and consumption habits, habits when using social media, etc.
- Factors related to the campaign carried out: goals, message, means and supports, temporary programming, relationship with other actions of the communication plan, etc.

Answer to Question 2. *In the case of brands with an image traditionally linked to other attributes, how can they reach consumers who are sensitive to sustainability?*

How to give value to the commitments assumed with society and the environment, even if they are below those of other sustainable cosmetic brands?

Based on the factors identified in the previous question, students can reflect about the causes explaining the lesser effectiveness of the campaigns conducted by 'conventional' companies and brands as well as try to identify alternative ways of communicating social and environmental commitments in these cases.

The question is adapted to the format of creative techniques such as brainstorming, so that students can make proposals in small groups and then discuss them as a whole class, under the guidance of the teacher. With this purpose, ideas about actions linked to the campaigns in social networks or other alternative communication actions (labels and certifications, sustainability reports, public relations, etc.) can be proposed.

Answer to Question 3. *Is it necessary to work on empowering consumers against greenwashing campaigns, or do they have their own resources to judge them appropriately?*

This is a question with a free response (there are no correct or incorrect answers). A general discussion can be set up in the whole class in which each student could contribute a personal opinion. It is important that students strongly argue their opinions towards the campaigns analysed in the case, whether they are favourable or unfavourable. After the previous round of interventions, the teacher can promote a debate with the focus on reflecting about the resources and knowledge of consumers when aiming to identify sustainable products in the market and to judge the promotional campaigns of companies that communicate their social and environmental commitments. It is also important to redirect the debate to discuss the scepticism of many consumers towards any campaign or communication action related to sustainability as well as about the potential causes of this attitude.

Answer to Question 4. *To what extent can a segmentation between millennial and centennial consumers be useful in the sustainable cosmetics sector? And in other contexts?*

This is also a question with a free answer. The previously generated debate at the whole class level can be continued, so that the students could think of examples of actions and communication campaigns linked to the sustainability of brands in which it can be useful to differentiate between consumers belonging to different generations.

To lead the discussion, the teacher can complement the information provided in the case with additional data regarding the characteristics linked to consumers of the millennial and centennial generations as, for example:

- Millennials: individuals showing a high use of new technologies (to the extent of juggling up to three devices at the same time), who were born in a time of economic boom, have experienced an overeducation process, and are used to job rotation every 3 years, tend to buy online, show a negative attitude towards advertising, are interested in brands, show an orientation towards individualism and little solidarity, are environmentally concerned but show little involvement, are more cautious and less idealistic than their elders, etc.

– Centennials: individuals with a great dependence on technology (to the extent of juggling up to five devices at the same time), who were born in a time of economic crisis, are short-term oriented, impatient and self-taught, tend to be self-employed and become entrepreneurs, prefer to buy in physical stores, are less consumerists, ask for inclusive and not stereotyped advertising, were more educated in issues related to equality or racial and cultural diversity, demand a greater environmental commitment to brands, etc.

Answer to Question 5. *What are the limitations of the positioning study presented in the case, and how could the situation be improved?*

When solved in advanced levels (by students in their last degree courses, in MBA seminars, etc.), the case can be used to introduce more complex notions about market research, by addressing issues such as the sample size and representativeness of the positioning study, the selection of the brands which were included, the choice of the analysis attributes which were considered in the maps presented, the analysis technique used, etc.

The intended purpose is that by working in small or large groups, students question the scope of the obtained results and their possibilities of generalisation, thus completing the reflection about the conclusions of the case presented and exploring further research opportunities.

References

Clinique. (2021). *Instagram*. Available from https://bit.ly/3iQpXRG (last accessed 15 December 2022).

Clinique. (2022). *Web site*. Available from https://bit.ly/3sS0BYG (last accessed 15 December 2022).

Composed Creative. (2019). *Natural + Sustainable Skincare for Millennials & Gen Z*. Available from https://s3.amazonaws.com/composedcreative_wp/wp-content/uploads/2019/04/19174221/Composed_Natural_Sustainable_Skincare_Report.pdf (last accessed 15 December 2022).

Cosmetics Europe. (2017). *Socio-economic development & environmental sustainability: the European cosmetics industry's contribution*. Available from https://bit.ly/3qP41um (last accessed 15 December 2022).

Dove. (2022a). *Instagram*. Available from https://bit.ly/3uJNx7U (last accessed 15 December 2022).

Dove. (2022b). *Web site*. Available from https://bit.ly/3yVil9v (last accessed 15 December 2022).

Eixarch, H., Wyness, L., & Sibanda, M. (2019). The regulation of personalized cosmetics in the EU. *Cosmetics, 6*(2), 29.

Fortuny, C., Oliver-Solà, J., & Gasol, C. M. (2020). Cómo incorporar y evaluar la sostenibilidad en la industria cosmética. *Noticias de Cosmética y Perfumería, 375*, 4–9. Sociedad Española de Químicos Cosméticos. Available from https://bit.ly/3qQuQ1o (last accessed 15 December 2022).

Freshly Cosmetics. (2022a). *Instagram*. Available from https://bit.ly/3JXTxjT (last accessed 15 December 2022).

Freshly Cosmetics. (2022b). *Web site*. Available from https://bit.ly/38bdExl (last accessed 15 December 2022).

Keller, K. L. (1993). Conceptualizing, measuring, and managing customer-based brand equity. *Journal of Marketing, 57*(1), 1–22.

Keller, K. L. (2003). Brand synthesis: The multidimensionality of brand knowledge. *Journal of Consumer Research, 29*(4), 595–600.

L'Oréal. (2021). *Instagram*. Available from https://bit.ly/3PI2pxA (last accessed 15 December 2022).

L'Oréal. (2022). *Web site*. Available from https://bit.ly/3NvCAi3 (last accessed 15 December 2022).

Lush. (2022a). *Instagram*. Available from https://bit.ly/35qRrtZ (last accessed 15 December 2022).

Lush. (2022b). *Web site*. Available from https://bit.ly/3PBpSQJ (last accessed 15 December 2022).

Maybelline. (2022a). *Instagram*. Available from https://bit.ly/3IVSdge (last accessed 15 December 2022).

Maybelline. (2022b). *Web site*. Available from https://bit.ly/3yOUZT4 (last accessed 15 December 2022).

Shen, B., & Bissell, K. (2013). Social media, social me: A content analysis of beauty companies' use of Facebook in marketing and branding. *Journal of Promotion Management, 19*(5), 629–651.

Stanpa, Asociación Nacional de Perfumería y Cosmética. (2020). Estudio del mercado de perfumería y cosmética en España. *Resultados económicos de 2019 y perspectivas 2020*. Available from https://bit.ly/35r6qE5 (last accessed 15 December 2022).

Statista. (2022). *Cosmetics industry – Statistics & Facts*. Available from https://bit.ly/3NBrQiZ (last accessed 15 December 2022).

Swanepoel, C., Lye, A., & Rugimbana, R. (2009). Virally inspired: A review of the theory of viral stealth marketing. *Australasian Marketing Journal, 17*(1), 9–15.

Part V

Sustainability in/from Other Organizations

Responsible Consumption and Global University Sustainability Through the UI GreenMetric World University Rankings

Tania Blanco-Sánchez, María Victoria Carrillo-Durán, and María García-García

Learning Objectives
The following are the main learning objectives in this case:

- To reflect on the potential of universities as educators in responsible consumption and sustainability
- To identify the sustainable development policies of the best universities in the ranking within the context of the SDGs
- To highlight possible improvements in terms of responsible and sustainable management of University Social Responsibility

1 Introduction

The efficient management of our shared natural resources and the way we dispose of waste form part of the 2030 Agenda's SDGs and are becoming increasingly important. Universities, as leading educational centres and organisations globally, need to improve the way they consume, recycle and reduce waste. This is increasingly becoming an obligation imposed on them.

On a general level, according to Ulusoy (2016), responsible consumption is consumption that has less of a negative impact or more of a positive impact on the environment, society and the individual. This general concept includes specific practices that must be adopted by universities and all their stakeholders (teaching staff, students, management bodies, senior management, etc.), such as sustainable consumption that reduces the impact of our present actions on the future and ethical consumption that does not negatively affect others. It also involves promoting consumer citizenship, meaning that consumers prioritise and actively contribute to fair development. Finally, it involves socially responsible consumption, which

T. Blanco-Sánchez · M. V. Carrillo-Durán (✉) · M. García-García
University of Extremadura, Badajoz, Spain
e-mail: taniabs@unex.es; vicduran@unex.es; mgargar@unex.es

E. C. Bianchi et al. (eds.), *Responsible Consumption and Sustainability*, Springer Business Cases, https://doi.org/10.1007/978-3-031-30742-3_15

improves social problems, and green consumption, which highlights environmental problems above others.

In addition, the term sustainability also has a prominent place on the agenda of universities, which must therefore take into account their environmental and social impacts and align them with the SDGs. Sustainability is considered the result of economic profitability, respect for the environment and social responsibility (Johnson, 2009). Higher education plays a fundamental role in training professionals who, in the future, will lead society and be responsible for improving economic profitability, caring for and respecting the environment and showing responsible behaviour towards society. Universities, therefore, enhance sustainability in their communities by transforming their attitudes and promoting behaviour that supports sustainability. In this way, higher education institutions contribute to preparing individuals so that in their personal and professional lives they can help to resolve these problems, applying the most appropriate technologies, developing more efficient industrial processes, promoting the individual development of people and societies, promoting the development of less developed countries, etc.

This paper addresses responsible consumption and sustainability from a university social responsibility perspective, analysing the top 100 universities in the UI GreenMetric World University Rankings (2021). This ranking measures the implementation of the Sustainable Development Goals (SDGs) at universities and, therefore, assesses the current position and policies related to responsible consumption and sustainability at higher education institutions worldwide.

Given the growing importance of the responsible consumption policies of universities in the current context of sustainability, the UI GreenMetric looks at six variables directly related to the SDGs and highlights the responsible consumption of universities, resulting in a map of university social responsibility, an aspect that has become more important recently (Păunescu et al., 2017) as a management and image-enhancing tool for universities. However, as educational institutions, universities must adopt a critical stance and actively fulfil their responsibility (Hindiyeh et al., 2022) by implementing exemplary sustainability policies, which go beyond being a direct policy to attract students (Saleh et al., 2021) and instead become a cross-cutting approach that underpins all university activities and objectives in terms of consumption and sustainability.

2 Case Development

The UI GreenMetric World University Rankings group the SDGs from the 2030 Agenda into six major consumption blocks that are observed and measured: setting and infrastructure, energy and climate change, waste, water, transportation, and education and research. This creates an accurate snapshot of the current position in terms of the responsible consumption and sustainability of universities, as fundamental elements of university social responsibility.

Table 1 Summary of the main blocks in the ranking and their relationship with the SDGs

Blocks	SDGs involved	Description of the block
Waste	SDGs 3, 12, 14, 15 and 17	Looks at consumption from the perspective of waste prevention and reduction by the university community
Water	SDGs 6 and 17	Looks at the rational and sustainable consumption of water as a basic resource and tackles the risks related to its scarcity
Transportation	SDGs 11, 13 and 17	Looks at consumption through the establishing of transportation policies that limit the number of motorised vehicles on campus and encourage the use of buses and bicycles.
Energy and climate change	SDGs 7, 11, 13 and 17	Looks at energy consumption in terms of its impact on climate change. This is the indicator with the highest weighting in the ranking
Education and research	SDGs 1, 2, 3, 4, 5, 8, 9, 10, 13, 14, 15, 16 and 17	Looks at consumption through the number of sustainability-related courses offered or research funds dedicated to sustainability
Setting and infrastructure	SDGs 11, 12 and 17	Looks at consumption by establishing actions to improve environmental management in which the university community as a whole plays a leading role

Source: Prepared by the authors based on the UI GreenMetric (2022c)

This case study is based on the analysis of the top 100 universities in the UI GreenMetric World University Rankings, which measure the implementation of the SDGs by universities, grouping them into six large blocks (see Table 1).

From the analysis conducted, it can be seen that universities manage their consumption well, therefore demonstrating good environmental responsibility (Fig. 1) as an indicator of their level of university social responsibility.

In Fig. 1, the most notable aspects are consumption in terms of education and research, followed by water management, waste management and transportation.

Therefore, on the one hand, in the context of sustainable development, improving local waste management is becoming a global priority. From the analysis of the top 100 universities, it can be seen that universities invest a great deal of effort in implementing recycling programmes, organic waste treatment, inorganic waste treatment, recycling of toxic waste, wastewater disposal and policies to reduce the use of paper and plastic on campus.

On the other hand, sustainable transportation is an issue intrinsically linked to waste, as the more people there are on the move, the more waste is generated from these activities (Wardani et al., 2021). Therefore, this is one of the main challenges universities are facing due to the increasing environmental and social impacts of the daily movements of their population (Miralles-Guasch & Domene, 2010). Hence, the interest of these organisations in implementing sustainable transport plans, with transport policies that limit the number of motorised vehicles or encourage the use of buses and bicycles on campus.

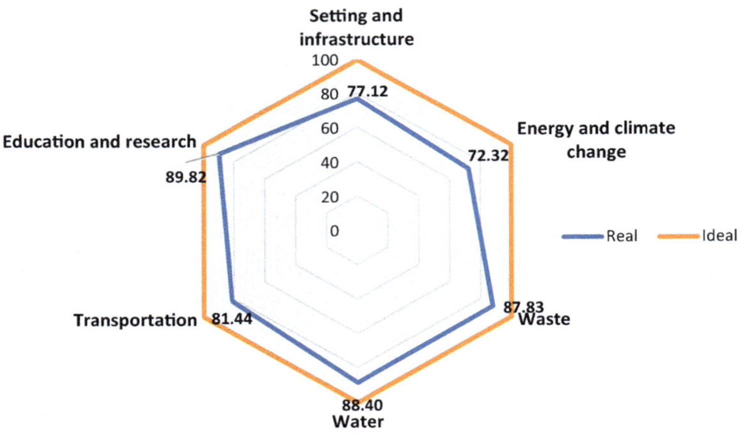

Fig. 1 Environmental responsibility management of the top 100 universities in the ranking Source: Prepared by the authors based on the UI GreenMetric (2022c)

In contrast, according to the ranking, universities should strengthen actions related to setting and infrastructure, energy and climate change (Fig. 1). However, many of the universities perform actions to improve environmental management on their campuses or in their faculties in which the entire university community plays a leading role. The goal is to encourage participating universities to provide more green spaces and protect the environment, as well as to develop sustainable energy sources. Moreover, the scarcity of energy leads universities to try to obtain clean energy by becoming increasingly self-sufficient. It also encourages them to use energy-efficient appliances, create smart buildings/building automation and take action in areas such as the policy on renewable energy use, total electricity use, energy conservation programmes, green building elements, climate change adaptation and mitigation programmes, greenhouse gas emission and carbon footprint reduction policy.

However, on a general level, the top 100 universities in the ranking have a behavioural focus on responsible and sustainable consumption and on economic and social issues related to sustainability. This can be seen in the main actions they perform within the university itself, as shown in Table 2.

It is also important to highlight that many of the universities in the ranking perform actions related to the environment and sustainable consumption through a certified quality system, such as the ISO standards, mainly ISO 9001 (Quality Management System) and ISO 14001 (Environmental Management), in which Wageningen University & Research and Nottingham Trent University stand out. The latter obtained ISO 14001 certification in 2012, recognising the university's work to ensure environmental management.

Specifically, there are some universities that stand out in each of the six blocks represented in the ranking (Fig. 2) and that establish action programmes closely

Table 2 Summary of the main environmental actions of the top 100 universities in the ranking

Aspect	SDGs	Main actions
Education and research (ED)	1, 2, 3, 4, 5, 8, 9, 10, 13, 14, 15, 16, 17	• Academic publications on sustainability • Sustainability-related cultural events and activities • Community sustainability services organised by and/or involving students • Sustainability Reports
Water (WR)	6, 17	• Recycling of water • Use of water-saving appliances
Waste (WS)	3, 12, 14, 15, 17	• Recycling of university waste • Reducing the use of paper and plastic on campus • Treatment of organic, inorganic and toxic waste
Transportation (TR)	11, 13, 17	• Use of zero-emission vehicles • Initiatives to reduce private vehicles on campus
Setting and infrastructure (SI)	11, 12, 17	• University budget for sustainability efforts • Total area of the campus is covered with planted vegetation • Conservation: plants, animals and wildlife, genetic resources for food and agriculture secured in medium- to long-term conservation facilities
Energy and climate change (CE)	7, 11, 13, 17	• Use of energy-efficient appliances • Use of renewable energy sources on campus • Implementation of smart buildings

Source: Prepared by the authors based on the UI GreenMetric (2022c)

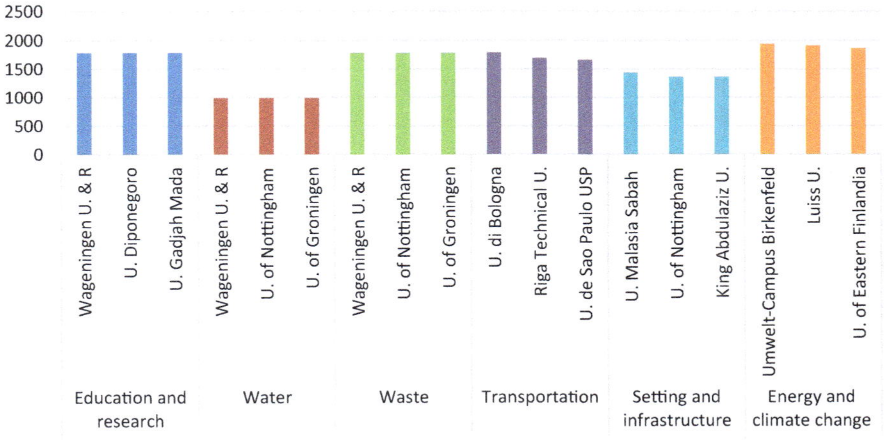

Fig. 2 Map of the environmental responsibility of the best universities in the ranking by each block. Source: Prepared by the authors based on the UI GreenMetric (2022c)

linked to each variable measured, serving as an example for others, with the confidence that their good practices have led them to the top positions in terms of responsible consumption and sustainability.

The following is a sample of the policies that are significant in the context of the universities achieving the best positions in each block.

2.1 Education and Research

Wageningen University & Research (Netherlands) considers sustainability to be an important aspect of research, education and value creation. That is why it applies the 'living lab' methodology, bringing sustainability closer to the university community by making the campus the venue of an annual week-long festival to commemorate National Sustainability Day ('De Dag van de Duurzaamheid'). During this week, different sustainability topics are covered, and a wide range of sustainable initiatives on and around the campus are showcased. The programme is different each year, but it always seeks a balance between academic and leisure activities, both for students and staff.

With initiatives such as this one, universities create a sustainable development culture through (Dzimińska et al., 2020): (1) teaching that covers sustainability challenges, (2) research based on real-world problems and (3) engagement with individuals and institutions.

2.2 Water

Universities are aware of the importance and scarcity of water, which is why they prioritise the rational and sustainable consumption of water as a resource that is a basic necessity. As a result, almost all universities have water conservation and recycling programmes, use water-saving appliances, consume treated water and control water pollution.

In addition to the above actions, adopted by almost all universities, Nottingham Trent University (UK) collects rainwater and reuses it in the university, an initiative also used by Wageningen University & Research (Netherlands).

2.3 Waste

In this block, over 20 years ago, Wageningen University & Research (Netherlands) began to manage its raw materials and waste with a 'circularity vision', with which it halved the waste generated through actions such as sharing products (one phone with two SIM cards, one private and one for work), using multifunctional products (printer-scanner-photocopier-fax) or refraining from using products (not using desk-top computers in the practical rooms). Furthermore, as part of this circularity, it uses products for a longer period of time and reuse others for different purposes, either at the university itself or elsewhere. For example, it recycles coffee cups to turn them into toilet paper, transforms waste paper into toilet paper and invites the university community to bring their own mugs or use reusable coffee cups on campus. The university also does not use certain products that may be polluting as construction materials, décor, furniture or cleaning services and uses other responsible and sustainable consumption products. It has also created electronic ordering and invoicing procedures and eliminated paper.

2.4 Transportation

In this block, the Universita di Bologna (Italy) is particularly noteworthy, having implemented a range of measures to promote sustainable mobility and encourage the use of public transport as part of the Sustainable Multicampus Project. Some of the measures taken have included subsidising transport with passes for the students and staff at the university and the complete replacement of its fleet with electric vehicles. Along this same line, Wageningen University & Research (Netherlands) encourages car sharing among members of the community, has electric charging points for bicycles and electric scooters, has a bicycle programme for staff members and has created additional bicycle lanes and bicycle parking areas on campus.

2.5 Setting and Infrastructure

Universiti Malaysia Sabah (Malaysia) excels in the configuration of the campus and information on the infrastructures, creating green spaces, protecting the environment and using sustainable energy. Therefore, through the EcoCampus Transformation Plan, it performs activities such as the maintenance of recreational areas like beaches, the subsidising of sustainability-related activities and the creation of the EcoCampus Creativity Awards, which are given to internal organisations for promoting environmental sustainability. Another initiative for which this university stands out is EcoCampus Tourism, which promotes five tourist packages ('Gallery and Museum Tour', 'Nature Delight Tour', 'Architectural Heritage Tour', 'Sunset Tour' and 'Nature Education Tour') within the university itself and is based on the concepts of nature tourism and green tourism.

Outside the top three in this block, other universities stand out for implementing sustainability measures relating to the construction, maintenance or renovation of buildings, such as Wageningen University & Research (Netherlands), and in the regulation of newly constructed buildings, as well as in energy use, as in the case of the University of Groningen (Netherlands), which constructs buildings that are almost energy neutral.

2.6 Energy and Climate Change

With regard to emissions and the carbon footprint, several universities have adopted important measures. At Umwelt-Campus Birkenfeld (Germany), every technical facility on campus is designed with the aim of making it zero emission. To do this, it has implemented movement sensors, installed photovoltaic systems on the roofs, oriented the roofs to avoid elements that may cause shade, installed daylight regulation systems and has a central absorption cooling unit that cools the air, floors and walls through a cooling circuit. Its heating is provided by a wood-fired electricity power plant. The technology installed on campus, therefore, guarantees 100% coverage of its basic energy needs from renewable sources.

Another university outside the top three in this block, Wageningen University & Research (Netherlands), generates 100% of the energy it consumes through sustainable methods such as wind turbines, thermal storage and solar panels on campus. It also has the Shut the Hood project, which is focused on informing, motivating and encouraging both students and employees about the safe and sustainable use of energy.

2.6.1 Other Measures

Aside from the blocks measured in the UI GreenMetric World University Rankings, there are other measures taken by universities that provide them with information about the work they must do to improve responsible consumption and university sustainability in the context of their university social responsibility.

One such action that deserves special attention is the effort universities are making with regard to the food they offer on their campuses. Many studies have focused on how universities could support sustainable food systems.

Some universities such as Wageningen University & Research (Netherlands) are leading the way by offering food of which more than 50% is obtained from sustainable sources, and more than 40% of the beverages are organic and take animal welfare into account. Along this same line, Nottingham Trent University (UK) offers food products associated with sustainable agriculture, including animal welfare practices. For example, they only use free-range eggs, quality-assured meat and organic milk.

The university also has other initiatives linked to responsible consumption and sustainability. One of these is a Foodshare Allotment, which is a food growing project for members of the university community who are taught how to grow fruit and vegetables without pesticides. Some of the produce grown is given to the project participants and the rest is donated to local charities and community cafés. Another measure is the Fair-Trade Policy, and for this, it organises fair-trade activities and events each year in order to raise awareness in society, such as 'fair-trade afternoon teas', street food markets, desserts in the university's catering establishments, cake competitions and fair-trade days and debates. It also encourages its university community to perform green actions and engage in more sustainable and wellness behaviours with the 'NTU Green Rewards Programme'. To participate in the programme, users simply need to download a free app where they register their sustainable actions and monthly challenges, which are awarded a number of points. Each month the top two staff members and top two students earn a £20 voucher, which they can redeem in selected shops or donate in their name to a local charity. More than 3,500 university employees and students have participated in the Green

Rewards, and as a result, more than 200,000 sustainability actions have been carried out.

3 Conclusions

It is clear that the 2030 Agenda and the SDGs are forming part of the strategic plans of universities. The incorporation of the SDGs into university governance, management and culture will enable universities to position themselves as leaders of the social transformation required for respectful development. They become exponents of university social responsibility managed through a series of policies on responsible consumption and sustainability, among others.

It is, therefore, no surprise that the best universities in the UI GreenMetric World University Rankings develop environmental sustainability strategies, in which responsible consumption and sustainability are key priorities, two or three times a year. These plans are called different things, such as the Integrated Sustainability Programme, Strategic Sustainability Plan, Strategic Plan for Sustainable and Environmental Development, Environmental Roadmap, etc.

Moreover, the dissemination of the actions performed is fundamental so that universities can demonstrate to their community that they are committed to sustainable development. The universities regularly (normally annually) draw up a report on sustainable practices and responsible consumption. This report not only contains the policies for sustainable practices but also the responsible consumption and sustainability actions carried out, as well as the achievements linked to these, such as education, research or sustainable investments. The UI GreenMetric World University Rankings itself has an indicator, within the Education and Research category, entitled Sustainability Report, which gives points to the universities that produce this report.

However, within the large blocks contemplated in the ranking, there are some areas of priority attention for universities. Undoubtedly, the consumption of education and research, in the form of courses and projects, are the areas in which universities invest most effort. Having syllabuses that include courses focused on promoting sustainability is a growing trend, as is the fact of considering responsibility as a cross-cutting issues for all the courses offered.

Mobility is another aspect that universities are quite concerned about, encouraging the community to use transport options with a lower environmental impact or promoting public transport. Likewise, water and waste management are areas to which higher education institutions pay a lot of attention.

However, caring for the environment and investing in environmentally friendly infrastructures are a challenge for universities. The use of renewable energies and the fight against climate change also present major challenges.

Universities focus their efforts on teaching and disseminating sustainability and responsible practices, above all among students, although they should not forget the institution's other stakeholders who could be more involved in sustainable management and consumption policies, such as the teaching staff, administrative staff and senior management, among others. As a result, it is necessary to understand that university social responsibility strategies should not only be based on good consumption and sustainability practices within the university itself but instead should also be implemented through actions forming part of broader action programmes in which other agents such as local, regional or central governments are often involved.

With the consumption and sustainability actions described throughout this paper, universities seek a balance between scientific, social and economic interests, the pillars on which their university social responsibility can be based, as a unifying asset for all these actions and a letter of introduction for the institutions internally and externally.

4 Discussion Questions

1. What do you think are the main difficulties for the development of responsible consumption and sustainability by universities?
2. What would you say are the main advantages and disadvantages of these policies for different university stakeholders? Do you think that universities are taking into consideration different stakeholders?
3. In the context of responsible consumption and sustainability, would you say that the most important universities in the ranking are doing well with being responsible?
4. What are the benefits to a university undergoing this evaluation process? What are the advantages and disadvantages of this ranking?
5. Do you consider that university social responsibility depends more on some policies than others (thinking about those applied by universities in the different aspects/variables considered in the ranking)?
6. At your university, what CSR actions do you find most interesting? What other actions could your university carry out to improve its policies on responsible consumption and sustainability?

Acknowledgements This research is funded by the Junta de Extremadura (Consejería de Economía, Ciencia y Agenda Digital) and by the European Union 'European Regional Development Fund. A way of shaping Europe', GR21028.

JUNTA DE EXTREMADURA

Consejería de Economía, Ciencia y Agenda Digital

Unión Europea

European Regional Development Fund
A way of doing Europe

Appendix: Teaching Notes for the Case

Synopsis

This case is based on the UI GreenMetric World University Rankings which measure the implementation of the Sustainable Development Goals (SDGs) at universities and assess the current position and policies related to responsible consumption and sustainability at higher education institutions.

The ranking groups the SDGs into six major blocks: setting and infrastructure, energy and climate change, waste, water, transportation, and education and research, offering an accurate snapshot of the current situation of universities in terms of meeting the SDGs from a university social responsibility perspective.

This paper takes as a reference the first 100 cases of universities appearing in the UI GreenMetric World University Rankings (2021) and seeks to identify the universities that implement best practice in responsible consumption and sustainability in each block, exemplify some of them and promote reflection on and better designs for university social responsibility strategies.

Potential Audience and Instructor's Material

Its contents are suitable for studying responsible consumption and sustainability, through a real and classic example. Specifically, the case is based on the UI GreenMetric World University Rankings (2021). It could, therefore, be useful for undergraduate or graduate courses in economics, business, marketing or advertising, among others.

Learning Objectives

The case is structured to meet the following objectives:

- To reflect on the potential of universities as educators in responsible consumption and sustainability
- To identify the sustainable development policies of the best universities in the ranking within the context of the SDGs
- To highlight possible improvements in terms of responsible and sustainable management of university social responsibility

Time Frame for Class Discussion

This case can be taught in one session of 75 min.

Suggested Discussion Questions

1. What do you think are the main difficulties for the development of responsible consumption and sustainability by universities?
2. What would you say are the main advantages and disadvantages of these policies for different university stakeholders? Do you think that universities are taking into consideration different stakeholders?
3. In the context of responsible consumption and sustainability, would you say that the most important universities in the ranking are doing well with being responsible?
4. What are the benefits to a university undergoing this evaluation process? What are the advantages and disadvantages of this ranking?
5. Do you consider that university social responsibility depends more on some policies than others (thinking about those applied by universities in the different aspects/variables considered in the ranking)?
6. At your university, what CSR actions do you find most interesting? What other actions could your university carry out to improve its policies on responsible consumption and sustainability?

Analysis

Answer to Question 1. *What do you think are the main difficulties for the development of responsible consumption and sustainability by universities?*

Universities face difficulties such as the need for financial resources to fund these policies, over and above other needs. However, the presence of the universities in the ranking and the overall results of the analysis show a high level of interest in these sustainability and responsible consumption policies.

There are also problems related to creating and implementing costly infrastructures, not only in terms of money but also in terms of time and other resources needed to put them in place efficiently.

The need to plan actions in this area is based on the conviction of management, so it is no surprise that the best universities in the UI GreenMetric World University Ranking develop environmental sustainability strategies, in which responsible consumption and sustainability are key priorities, two or three times a year. These plans are called different things, such as the Integrated Sustainability Programme, Strategic Sustainability Plan, Strategic Plan for Sustainable and Environmental Development, Environmental Roadmap, etc.

However, within the large blocks included in the ranking, there are some areas of priority attention and others that are less so, due to the difficulty involved in reaching new milestones in these areas, such as caring for the environment and investing in environmentally friendly infrastructures. The use of renewable energies and the fight against climate change also present major challenges.

Answer to Question 2. *What would you say are the main advantages and disadvantages of these policies for different university stakeholders? Do you think that universities are taking into consideration different stakeholders?*

Universities focus their efforts on teaching and disseminating sustainability and responsible practices, above all among students, although they should not forget the institution's other stakeholders who could be more involved in sustainable management and consumption policies, such as the teaching staff, administrative staff and senior management, among others. As a result, it is necessary to understand that university social responsibility strategies should not only be based on good consumption and sustainability practices within the university itself but instead should also be implemented through actions forming part of broader action programmes in which other agents such as local, regional or central governments are often involved.

Answer to Question 3. *In the context of responsible consumption and sustainability, would you say that the most important universities in the ranking are doing well with being responsible?*

From the analysis conducted, it can be seen that universities manage their consumption well, therefore demonstrating good environmental responsibility as an indicator of their level of university social responsibility.

The most notable aspects are consumption in terms of education and research, followed by water management, waste management and transportation.

In contrast, according to the ranking, universities should strengthen actions related to setting and infrastructure, energy and climate change. However, many universities perform improvement actions.

On a general level, the top 100 universities in the ranking have a behavioural focus on responsible and sustainable consumption and on economic and social issues related to sustainability. This can be seen in the main actions they perform within the university itself and their direct connection to the main SDGs.

Summary of the main environmental actions of the top one hundred universities in the ranking

Aspect	SDGs	Main actions

(continued)

Aspect	SDGs	Main actions
Education and research (ED)	1, 2, 3, 4, 5, 8, 9, 10, 13, 14, 15, 16, 17	• Academic publications on sustainability • Sustainability-related cultural events and activities • Community sustainability services organised by and/or involving students • Sustainability Reports
Water (WR)	6, 17	• Recycling of water • Use of water-saving appliances
Waste (WS)	3, 12, 14, 15, 17	• Recycling of university waste • Reducing the use of paper and plastic on campus • Treatment of organic, inorganic and toxic waste
Transportation (TR)	11, 13, 17	• Use of zero-emission vehicles • Initiatives to reduce private vehicles on campus
Setting and infrastructure (SI)	11, 12, 17	• University budget for sustainability efforts • Total area of the campus is covered with planted vegetation • Conservation: plants, animals and wildlife, genetic resources for food and agriculture secured in medium- to long-term conservation facilities
Energy and climate change (CE)	7, 11, 13, 17	• Use of energy-efficient appliances • Use of renewable energy sources on campus • Implementation of smart buildings

Source: Prepared by the authors based on the UI GreenMetric (2022c)

Answer to Question 4. *What are the benefits to a University undergoing this evaluation process? What are the advantages and disadvantages of this ranking?*

The main advantages to be in this ranking are fundamentally the following:

1. The incorporation of the SDGs into university governance, management and culture will enable universities to position themselves as leaders of the social transformation required for respectful development which will turn them into exponents of university social responsibility. With the consumption and sustainability actions described throughout this paper, universities seek a balance between scientific, social and economic interests, the pillars on which their university social responsibility can be based, as a unifying asset for all these actions and a letter of introduction for the institutions internally and externally.
2. The ranking increases the visibility of the actions performed, presenting and justifying their commitment to sustainable development. The universities regularly (normally annually) draw up a report on sustainable practices and responsible consumption. The UI GreenMetric World University Rankings itself has an indicator, within the education and research category, entitled Sustainability Report, which gives points to the universities that produce this report.
3. Moreover, being among the top positions in the ranking is becoming a factor affecting the decisions made by current and potential students and not only those made by students but also those of the institutions that finance research projects

and may take into account this increasingly relevant indicator when granting funds that allow universities to develop and establish their status.

However, appearing in the ranking forces universities to:

1. Be extremely rigorous and consistent in all the actions they perform, as any discrepancy between what they do and the message they send out about what they claim to do could cause serious damage to their reputation.
2. International rankings use universal criteria that do not take into consideration the particular circumstances of each institution. The national situation and government policies are not taken into account, so the weighting established may require actions that are not promoted locally.

Answer to Question 5. *Do you consider that university social responsibility depends more on some policies than others (thinking about those applied by universities in the different aspects/variables considered in the ranking?*

Given the growing importance of the responsible consumption policies of universities in the current context of sustainability, the UI GreenMetric looks at six variables directly related to the SDGs (waste, water, transportation, energy and climate change, education and research, setting and infrastructure) and explains the responsible consumption of universities, resulting in a map of university social responsibility, an aspect that has become more important as a management and image-enhancing tool for universities.

University social responsibility is an intangible asset of great value that must be implemented across the six major areas described and evaluated in the ranking. However, its management does not only end with the implementation of responsible consumption policies and actions in these six areas but also requires efficient planning and communication. As seen above, on the one hand, planning is taking place in the universities that reach the top positions in the ranking, and, on the other, appearing in this ranking, among other promotional actions, is a way of declaring to society what they believe in, saying what they do that goes above and beyond what they are obliged to do.

Answer to Question 6. *At your university, what CSR actions do you find most interesting? What other actions could your university carry out to improve its policies on responsible consumption and sustainability?*

The idea is that students compare the examples given here with the situation in their own university and/or country and evaluate the pros and cons of their situation based on what they have learned here with other examples. Therefore, this question will be answered depending on the student and the individual characteristics of each case and situation.

Suggested Assignment Questions

1. In the context of responsible consumption and sustainability, what are the most important universities in the ranking doing well to be responsible? What are they not doing well?

The most notable aspects are consumption in terms of education and research, followed by water management, waste management and transportation.

In contrast, according to the ranking, universities should strengthen actions related to setting and infrastructure, energy and climate change. However, many universities perform improvement actions.

Generally speaking, the main actions can be summarised as follows:

Education and research (ED).

- Academic publications on sustainability
- Sustainability-related cultural events and activities
- Community sustainability services organised by and/or involving students
- Sustainability Reports

Water (WR)

- Recycling of water
- Use of water-saving appliances

Waste (WS)

- Recycling of university waste
- Reducing the use of paper and plastic on campus
- Treatment of organic, inorganic and toxic waste

Transportation (TR)

- Use of zero-emission vehicles
- Initiatives to reduce private vehicles on campus
- Setting and infrastructure (SI)
- University budget for sustainability efforts
- Total area of the campus covered with planted vegetation
- Conservation: plants, animals and wildlife, genetic resources for food and agriculture secured in medium- to long-term conservation facilities

Energy and climate change (CE)

- Use of energy-efficient appliances
- Use of renewable energy sources on campus
- Implementation of smart buildings

2. What are the benefits to a university undergoing this evaluation process?

1. To position universities as leaders of the social transformation required for respectful development which will turn them into exponents of university social responsibility. With the consumption and sustainability actions, universities seek a balance between scientific, social and economic interests, the pillars on which their university social responsibility can be based, as a unifying asset for all these actions, and a letter of introduction for the institutions internally and externally.
2. The ranking increases the visibility of the actions performed, demonstrating their commitment to sustainable development. The UI GreenMetric World University Rankings itself has an indicator, within the education and research category, entitled Sustainability Report, which gives points to the universities that produce this report.
3. Presence among the top positions in the ranking is becoming a factor affecting the decisions made by current and potential students and not only those made by students but also those of the institutions that finance research projects and may take into account this increasingly relevant indicator when granting funds that allow universities to develop and establish their status.

This research is funded by the Junta de Extremadura (Consejería de Economía, Ciencia y Agenda Digital) and by the European Union 'European Regional Development Fund. A way of shaping Europe', GR21028.

JUNTA DE EXTREMADURA

Consejería de Economía, Ciencia y Agenda Digital

Unión Europea

European Regional Development Fund

A way of doing Europe

Appendix I. UI GreenMetric World University Rankings

UI GreenMetric (2022a). Overall Rankings 2021. *https://greenmetric.ui.ac.id/rankings/overall-rankings-2021*

Appendix II. UI GreenMetric World University Rankings Questionnaire
UI GreenMetric (2022b). Questionnaire. https://greenmetric.ui.ac.id/publications/questionnaire

Appendix III. Indicators and categories suggested for use in UI GreenMetric World University Rankings
Guidelines 2021 – English. https://greenmetric.ui.ac.id/publications/guidelines/2021/english

References

Dzimińska, M., Fijałkowska, J., & Sułkowski, Ł. (2020). A conceptual model proposal: Universities as culture change agents for sustainable development. *Sustainability, 12*(11), 4635. https://doi.org/10.3390/su12114635

Hindiyeh, M., Jaradat, M., Albatayneh, A., Alabdellat, B., Al-Mitwali, Y., & Hammad, B. (2022). Sustainable Green University: Waste Auditing, German Jordanian University as a Case Study. *Frontiers in Built Environment, 8*, 884656. https://doi.org/10.3389/fbuil.2022.884656

Johnson, R. L. (2009). Organizational motivations for going green or profitability versus sustainability. *The Business Review, 13*(1), 22–28.

Miralles-Guasch, C., & Domene, E. (2010). Sustainable transport challenges in a suburban university: The case of the Autonomous University of Barcelona. *Transport Policy, 17*, 454–463.

Păunescu, C., Drăgan, D., & Găucă, O. (2017). Examining obligations to society for QS Stars best ranked universities in social responsibility. *Management & Marketing, 12*(4), 551–570. https://doi.org/10.1515/mmcks-2017-0033

Saleh, M. S. M., Kassim, N. M., Tukur, N. A., Mukhiar, S. N. S., & Balaraman, R. A. (2021). Sustainable universities as brand marketing for universities: A Case of Universiti Sains Malaysia. *Media Watch, 12*(1), 127–148. https://doi.org/10.15655/mw_2021_v12i1_205463

UI GreenMetric. (2022a). *Overall Rankings 2021.* https://greenmetric.ui.ac.id/rankings/overall-rankings-2021

UI GreenMetric. (2022b). *Questionnaire 2021.* https://greenmetric.ui.ac.id/publications/questionnaire

UI GreenMetric. (2022c). *Guidelines 2021 English.* https://greenmetric.ui.ac.id/publications/guidelines/2021/english

Ulusoy, E. (2016). Experiential responsible consumption. *Journal of Business Research, 69*(1), 284–297.

Wardani, W. K. et al. (2021). *IOP Conf. Ser.: Mater. Sci. Eng.* 1098 052046.

Understanding the Evolution of Organizations Towards a Sustainable Business Model Within the Framework of the SDGs: The Case of Laboratorio de Hemoderivados-UNC

Natacha Beltrán, Sergio Oviedo, Carina Dolores Oliva, Maria Angel Gomez-Auad, Juan Manuel Benegas-Prado Loyo, Mariana Mendoza, and Paulo Alejandro Morales

Learning Objectives

1. To contribute to the understanding of responsible production as a relevant dimension of the sustainability policy of an industrial organization in the pharmaceutical sector
2. To describe the approach to SDG 12 (responsible production and consumption) by an industrial organization inserted in a university institution
3. To analyse valuable initiatives of the organization to contribute to the sustainable development and human quality of the community
4. To assess the evolution and impact of the sustainable development actions carried out by the organization

N. Beltrán (✉) · S. Oviedo
Facultad de Ciencias Económicas, Universidad Nacional de Córdoba, Córdoba, Argentina

Laboratorio de Hemoderivados, Universidad Nacional de Córdoba, Córdoba, Argentina
e-mail: natacha.beltran@unc.edu.ar; sergio.oviedo@unc.edu.ar; 0732148@ucc.edu.ar

C. D. Oliva
Facultad de Ciencias Económicas, Universidad Nacional de Córdoba, Córdoba, Argentina

Universidad Blas Pascal, Córdoba, Argentina
e-mail: carina.oliva@unc.edu.ar; caroliva@ubp.edu.ar

M. A. Gomez-Auad · M. Mendoza · P. A. Morales
Laboratorio de Hemoderivados, Universidad Nacional de Córdoba, Córdoba, Argentina
e-mail: maria.gomez.auad@unc.edu.ar; mariana.mendoza@unc.edu.ar; alejandro.morales@unc.edu.ar

J. M. Benegas-Prado Loyo
Facultad de Ciencias Económicas, Universidad Nacional de Córdoba, Córdoba, Argentina
e-mail: juan.benegas@unc.edu.ar

E. C. Bianchi et al. (eds.), *Responsible Consumption and Sustainability*, Springer Business Cases, https://doi.org/10.1007/978-3-031-30742-3_16

1 Introduction

1.1 Towards Responsible and Sustainable Business Models

Innovation that takes place in the public sphere is transcendental for the community at large (Agarwal et al., 2021), although it experiences great challenges linked to resources, the capacity to innovate, and the generation of value that reaches the community (Hartley, 2006). Among the possible innovations, the business model deserves special attention as it is relevant to the creation of value for citizens, industry and government, where the different priorities presented by stakeholders deserve careful matching in order to meet diverse expectations.

A business model 'describes the logic of how an organization creates, delivers and captures value' (Osterwalder & Pigneur, 2010). Traditional business models often face challenges linked to the prioritization of monetary value at the cost of overexploitation of natural resources, jeopardizing their sustainability in the public interest. It becomes necessary to think of models that allow guaranteeing the interests of multiple stakeholders, benefiting the society and the environment and generating both monetary and nonmonetary value (Pan et al., 2022).

With the adoption of the 2030 Agenda for sustainable development by the United Nations, and its 17 SDGs, the concept of sustainability has regained vigour, as well as the sustainable business model and innovation in it is gaining momentum.

Among the various social organizations called upon to contribute to the achievement of the 2030 Agenda, universities stand out. They have the 'social responsibility' to promote competencies for sustainable development, not only among their students but also among the community (Franco et al., 2019). Universities play a great role, through linking and raising awareness in society and becoming living experimentation environments for sustainability (Ferguson & Roofe, 2020).

The adoption of sustainable business practices and models can be motivated by various factors, including the existence of competitive opportunities, the need for compliance and pressure from various stakeholders. Several theories have tried to explain the reason for the incorporation of such practices in both public and private organizations; among them, we can highlight the Legitimacy Theory, the Stakeholder Theory and the Signalling Theory (Hahn & Kühnen, 2013).

After having introduced the subject matter of this analysis, the case of the 'Laboratorio de Hemoderivados' (Hemoderivados or LH) of the Universidad Nacional de Córdoba (UNC) will be presented as an organization that assumes several roles, as a public pharmaceutical company under university management committed with sustainability management in the framework of the SDGs. Therefore, first, the context of the pharmaceutical industry as an actor committed to the SDGs and various sustainable practices will be introduced, to then address the innovation of LH's sustainable business model in recent years and finally to show how its different sustainable actions are reflected in the organization's sustainability report.

2 The Pharmaceutical Industry: A Stakeholder Committed to the SDGs

2.1 Responsible Consumption and Sustainability: Its Link with the Pharmaceutical Industry

Since the introduction of the nonfinancial reporting mandate for commercial entities by government authorities in different countries, the manufacturing landscape in the pharmaceutical industry has changed. The regulatory requirement from health authorities has been joined by the comprehensive requirement comprising a commitment to sustainable development. This situation is linked to the emergence of a new type of consumer, aware of their purchasing actions, with different values and concerned about their actions in the market (Crane & Matten, 2004; Izquierdo Maldonado et al., 2019).

Minimizing the effects on the environment is both good business and good corporate social responsibility (CSR), generating the introduction of sustainability practices in the development of new delivery systems, new products, greener manufacturing methods, reduction of water consumption and waste management that present a lower environmental risk and greater attention to this issue (Pérez Espinoza et al., 2016).

According to a study conducted in 2021 by Pharmaceutical Technology magazine on the sustainability areas that should be addressed by the pharmaceutical sector as a priority, it was obtained that out of 434 respondents from the pharmaceutical industry, 43% indicated environmental sustainability as the highest priority, 31% indicated social sustainability and 26% indicated governance sustainability (Fig. 1).

Among the priority aspects of environmental sustainability, 52% said that addressing climate change was a priority, and the rest indicated pollution (32%), natural resources (11%) and biodiversity (5%). When asked about social sustainability, the aspects of sustainability, health and safety were the most pressing with 49%, followed by human rights with 25% and community impact with 16%, while diversity and inclusion represented 9% (Pharmaceutical Technology News 1, 2021) (Fig. 2).

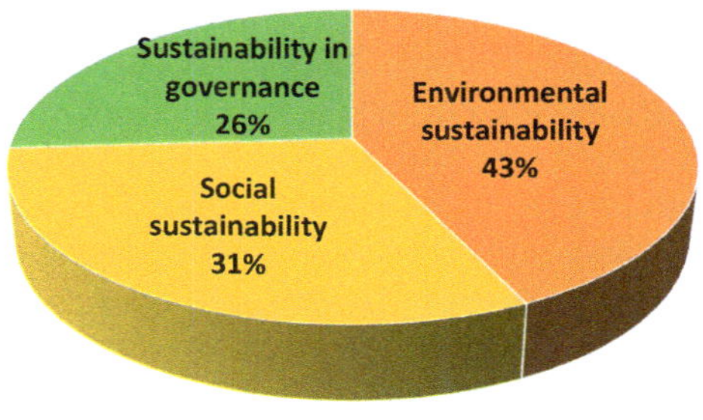

Fig. 1 Sustainability areas of greatest impact and need for development in the pharmaceutical industry Source: Adapted from Pharmaceutical Technology News 1 (2021)

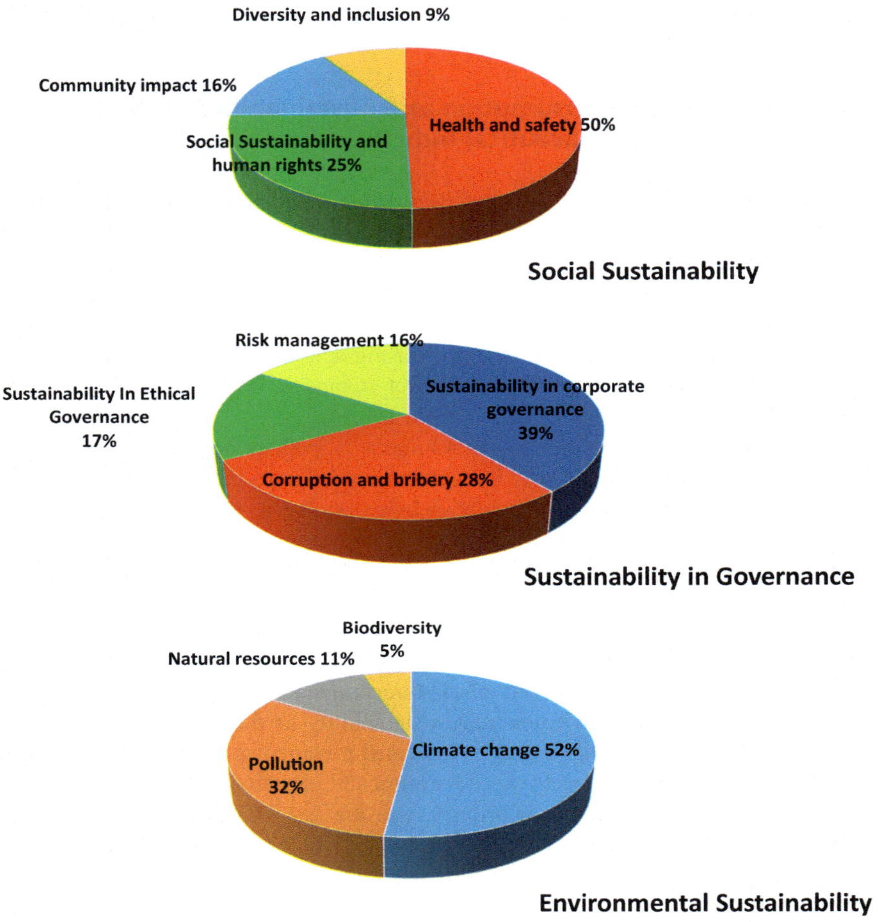

Fig. 2 Disaggregation of priority sustainability issues in the pharmaceutical industry Source: Adapted from Pharmaceutical Technology News 1 (2021)

2.2 Implications of the Pharmaceutical Industry and Its Processes in Sustainable Development

The link between environmental and social sustainability with human health is clear and becomes a priority in the strategies of the pharmaceutical industry (Neto, 2012).

The central idea revolves around the way of doing business and the influence of business on societies and the goals of sustainable development. While pharmaceutical companies are favoured by sustainability policies, which allow them to obtain long-term financial benefits, they open the doors to new markets and gain advantages over the competition. One of the strategies adopted is the creation of shared value (Vilariño & Bigorito, 2014), which allows recognizing that damaged or deprived societies create internal costs for companies, such as expenses in energy or raw materials, accidents and training to compensate for the lack of education. Investment in quality education does not necessarily increase costs, as firms can innovate through the use of new technologies, operational processes and management, thus increasing their productivity and expanding their markets (Veleva et al., 2003; Pharmaceutical Technology News 2, 2021).

Environmental sustainability represents another focus of attention in the sector, especially in terms of cleaner production, green supply chain, green materials and sustainable management of human assets (Weadon, 2010; Glavič, 2021; Rajagopal Ramachandran, 2017).

2.3 Trend of Practices Related to Sustainability

The pharmaceutical industry, insofar as it works to improve health and quality of life and promotes an economic model based on skilled employment, research, development and innovation, constitutes one of the key sectors on a global scale when it comes to making the Sustainable Development Goals a reality (Milanesi et al., 2020; Belkhir & Elmeligi, 2019).

Goal 12 Responsible production and consumption: This goal, as reflected in its targets 12.4, 12.5 and 12.8, integrates the actions of the pharmaceutical industry, insofar as it focuses on the recycling and environmental treatment of packaging and medicine waste and on people having relevant information and knowledge from the perspective of the safety, efficacy and final treatment of expired medicines.

3 The 'Laboratorio de Hemoderivados': A Sustainable Business Model

The 'Laboratorio de Hemoderivados' (LH) of the Universidad Nacional de Córdoba is an Argentine, public, non-profit, university-managed, industrial pharmaceutical organization. It has three production lines: blood-derived medicines, small-volume injectable drugs and human bone tissue for therapeutic use in traumatology and dentistry.

LH is an economically and financially self-managed state-owned company, with its income coming from the commercialization of products. By conceiving the drug as a social good, its policy has an impact on the substitution of imports of drugs of high therapeutic value and on the regulation of market prices. LH is a strategic player in the Argentine health system, as a producer of essential, orphan and strategic drugs and medical products (e.g. calcium EDTA, hematin, organ preservation fluid, bone implants).

Of its three business units, the plasma derivatives' unit is the main one, accounting for 90% of revenues. The plasma fractionation plant (the liquid part of blood, composed of proteins of therapeutic interest) is distinguished from the others in the world because it is the only one that belongs to a public university. Moreover, in the context of the Argentine health system, it is the only laboratory in the country that develops and manufactures blood-derived medicines. It supplies the domestic market and distributes its products in the countries of the region with which it has an agreement.

LH's business model is strongly affected by the fact that its main raw material, human plasma, is regulated by the National Blood Law 22.990, which prohibits the commercialization of blood and all its components.

Under these conditions, LH's value proposition to its main raw material suppliers—blood banks—is based on the exchange of its plasma surpluses for blood-derived medicines. Thus, instead of discarding the plasma as pathogenic waste—with the costs and environmental impact that this implies—it is sent to LH and converted into essential and strategic medicines for the health system.

A percentage of the drugs produced is sent to the member blood centres as remuneration, while the LH markets the rest, which allows it to cover operating costs and reinvest. Other forms of remuneration consist of training and modernization of equipment or support for publicity campaigns to encourage blood and plasma donation.

In this way, blood banks become strategic allies. This business model implies a strong commitment on the part of LH to the development of responsible citizenship, encouraging voluntary donation and thus contributing to the expansion of its productive capacity. This value proposition also structures the links with countries in the region such as Chile, Uruguay and Paraguay.

On the other hand, the business model is also characterized by marketing policies aimed at regulating market prices. The latter is key, not only for patients who have access to critical and very expensive treatments but also for various actors in the health system, such as social security, health institutions and the State itself. These policies also have an impact on the substitution of drug imports. Figure 3 shows the business model using the canvas proposed by Osterwalder (2010).

Thus, LH contributes to the health system, producing drugs of high therapeutic value, within the framework of a value chain that aims to be increasingly sustainable.

This can be reflected in the institutional mission of Hemoderivados:

> We are a non-profit public pharmaceutical laboratory, with a strong social commitment and with the pride of belonging to the Universidad Nacional de Córdoba; which seeks to improve people's quality of life by manufacturing accessible, high quality, safe and effective products, providing answers to critical demands in the national and regional health field.

In recent years, concern for transparency in management, labour practices, human rights and environmental issues have led to the development of SR&S policies linked to the SDGs.

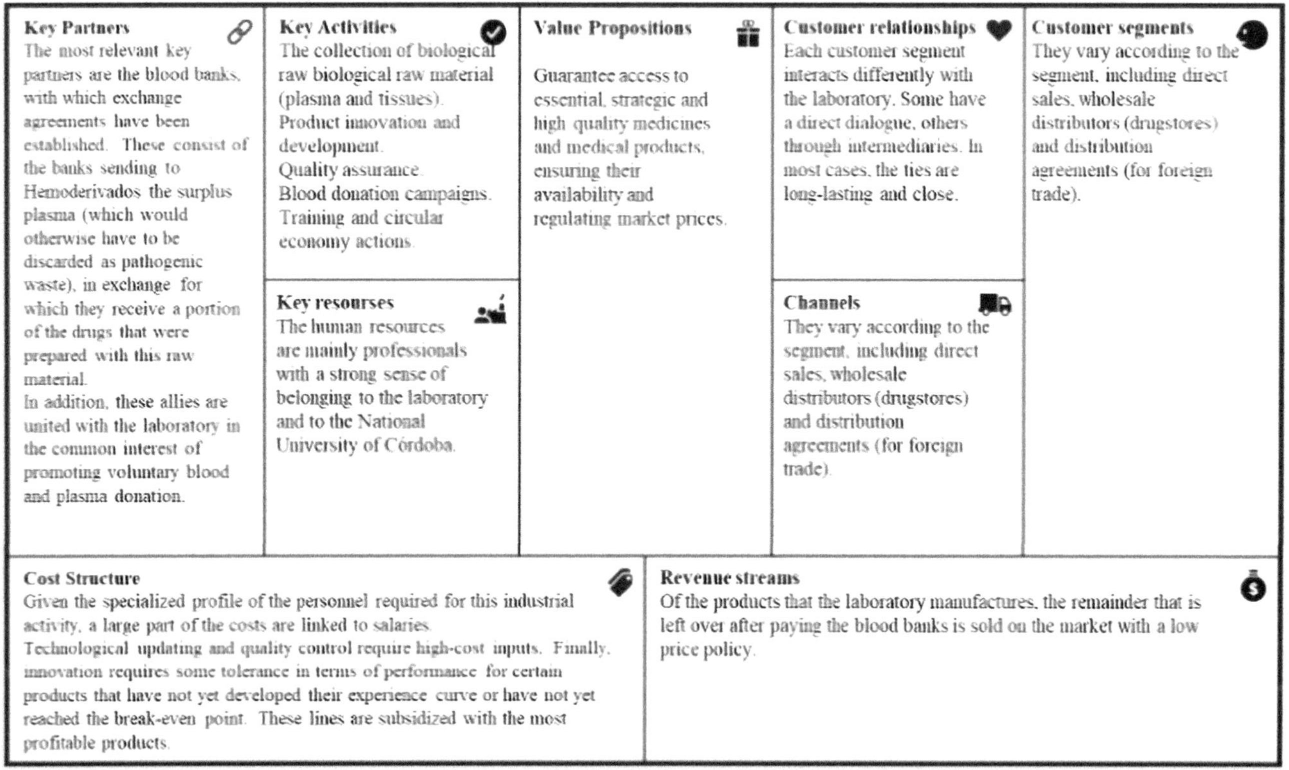

Fig. 3 Canvas of the 'Laboratorio de Hemoderivados' business model. Source: Own elaboration based on Osterwalder (2010)

4 Sustainability at LH: The Evolution Reflected in the Reports

The Laboratory, since its origin and creation, is an institution with strong principles of 'organizational social responsibility' (OSR), even when the concept as such was not defined as such and did not have the importance that is given to it nowadays. In this sense, it can be said that Hemoderivados is a pioneer institution in OSR, and this can be proven in its Mission and Vision and in each of the activities it carries out.

Its commitment to sustainable development is evidenced by the fact that it is a non-profit producer of strategic, unique and essential drugs; it intervenes in the market, guaranteeing access to and availability of these drugs; it regulates market prices; and it carries out social and health education and prevention actions. This commitment has gained strength more recently with the concern for management transparency, the development of proactive policies focused on people, human rights and environmental issues, which has led it to develop responsibility and sustainability policies linked to the SDGs and the 2030 Agenda.

In 2017, LH joined the United Nations Global Compact Network. And with this decision, the RSO was implemented, stimulating actions in relation to the SDGs.

The first sustainability reports of 2017 and 2018 were the initial kick-off in the path of compiling the triple impact activities, gathering all the relevant information to be captured in a document.

The change in the governance structure of the laboratory, which took place in 2019, strengthened the sustainability policy with the creation of the Sustainability Directorate and expanded the commitments made, also adopting the goals and plan of the Regional Conference on Higher Education (CRES) 2018–2028 and the Education for Sustainable Development (ESD) established by UNESCO and defined in the third CRES. Commitments were harmonized and links were strengthened with the SDG team of the University Extension Secretariat—of the University—to harmonize criteria and objectives with the development of SDGs and reports of the UNC.

In this new institutional framework, the Department of Integrated Waste and Effluent Management increased the activities that had been developed until then. Emphasis was placed on the management of all waste, significantly increasing all activities, looking at the path to be followed in an integrated and holistic manner.

Sustainability gave new impetus to the effluent treatment projects (Network 1 and Network 2), and progress was made in updating the injectable quality water generation plant with a better use of the resource and a 50% recovery of process water. The hydroalcoholic effluent recovery tank is currently being installed (Network 1), which in agreement with the company PORTA (ethanol producer) will allow the recovery of alcohol through a distillation process, thus establishing a reverse logistics system and complying with the requirements of the extended responsibility.

A customized station for processing PVC and blisters is in the process of being acquired. Through this process, it is expected that at least 60% of the waste will be processed in a sustainable manner to be reintroduced into the circular economy as raw material for another process. In addition, agreements were strengthened with the municipal entities in charge of collecting solid urban waste for reuse or recycling, thus benefiting the cooperatives associated with them.

In this new stage, alliances were also developed with the private sector for the processing of recyclable materials. In this way, a bidirectional economic flow is created.

In addition, and given that the LH is under the orbit of the UNC, it works as a team with those responsible for waste at the University. Training is currently underway to implement the Globally Harmonized System (GHS) of hazardous waste.

At the beginning of 2022, LH presented a new Sustainability Report that brings together what has been achieved in the 3-year period 2019–2021. The description of the main milestones and continuities embodied in these documents allows us to account for the evolution, both from conception and practice, of the concepts of sustainability, responsible consumption, etc.

4.1 Chronology of Significant Sustainability Initiatives

Regarding SDG 12 (Fig. 4), linked to responsible production and consumption, in 2017, LH continued to collaborate in import substitution as it has done since its origin; it added environmental commitment to social commitment and incorporated the concept of triple impact (social, economic, environmental). It also received a mention for productivity in the framework of the Industry Day Award, for the design and development of its new injectable quality water plant.

In 2018, meanwhile, LH incorporated guidelines for supplier management and audits of critical suppliers. Spending on local suppliers was also promoted. Focus was placed on energy consumption within the organization, an indicator was developed in this regard and work was done to reduce it. This involved reducing the energy requirements of products and services. Regarding water management, water

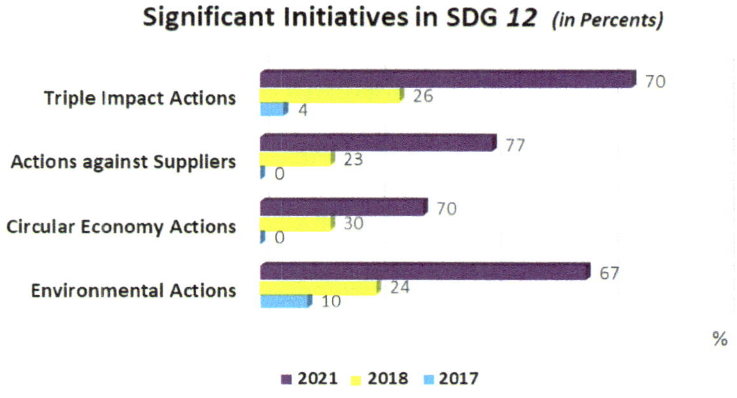

Fig. 4 SDG 12 major initiatives. Source: Prepared by the company

extraction by sources was implemented. We also worked on waste management by type and disposal method.

In the 2019–2021 period, within the framework of the Integral Waste and Effluent Management Program, work began with Green Centres of the Municipality of Córdoba for the processing of dry waste from the LH and cooperated with the Faculty of Agricultural Sciences for the management of precooking waste and the separation of foliage to be processed as compostable waste.

In addition, the grinder was refunctionalized for destruction of ampoules and vials, avoiding their discard as pathogens and achieving an average annual reduction of 500 kg between 2019 and 2020, and 7,921 Styrofoam boxes were sent for recycling to be used as construction filler (circular economy), between October 2019 and October 2021.

The Effluent Treatment Networks Project was launched, aimed at reusing supernatants, recovering dissolved alcohols and treating effluents based on their characteristics.

As regards materials and supplies, the Company began to use HFC refrigerant gases (hydrofluorocarbons, R404a, R410a), i.e. gases without chlorine molecules that do not damage the ozone layer, for refrigeration equipment. In addition, a sustainable purchasing project was initiated based on the principles of Good Pharmaceutical Practices and Sustainable Development. In order to optimize fuel consumption and vehicle fleet performance, LH began to use the 'YPF en Ruta' system. In this way, purchases were unified and routes were optimized by reducing idle routes. The increasing implementation of digital brochures reduced the use of paper and printing.

In terms of energy consumption, low-consumption lights were replaced with 60x60 LEDs, and the Energy Diagnosis program was initiated to implement an energy management system (SGEn) and develop alternative sources of energy generation.

In 2019, LH was distinguished by the United Nations with the Public Service of Excellence Award in the category 'Ensuring integrated approaches in the Public Sector'. In this way, its performance as a social, synergistic and sustainable business model was highlighted, oriented to the public production of medicines to ensure their availability and access to the population of Argentina and Latin American countries. That same year, it was awarded by INNDESA in the category Innovations that produce positive foreign exchange balances for Argentina.

In the context of the SARS COVID-19 pandemic, LH became the main supplier of dexamethasone for the Argentine State.

In this period, it also began to participate in the 'Follow-up Program for Sustainable Development Goals (SDGs) and the UN Agenda 2030', coordinated by the UNC University Extension Secretariat.

4.2 A Focus on Waste Treatment

In the compared period, work was carried out on the commitments assumed in environmental matters, advancing in issues related to the development of circular economy systems integrated to the productive and commercial circuit and in the search for solutions to the management of energy, waste and effluents, from infrastructure to the redesign of the business strategy integrating activities aimed at improving the quality of life of the community and the environment.

Figure 5 shows the evolution of waste management and its relative weight according to its classification. There is a clear increase in the volume of recyclable waste (5000 kg in 2017, 22,876 kg in 2021) to the detriment of pathogenic waste; this behaviour is coincident with the implementation of the integrated waste management program.

Figure 6 shows in detail the influence of the management system and sustainable policies from the decrease of pathogenic waste, from 2017 to date the annual weight decreased 1,481 kg per year, i.e. 1.78%, with an accumulated 7.19% to 2021.

Table 1 summarizes a SWOT analysis on the organization focused on the areas to consider according to the scope of sustainability.

The corporate sustainability model shown is intended to be the basis for the case analysis, and to recognize the factors (inputs) that impact on the company's process, and how from the leadership and sustainability policies (strategy, structure, systems and programs), the effects on stakeholders and sustainability indicators can be measured and evaluated, as well as the impacts on performance.

5 Conclusions

In this case, the main milestones in the evolution of the 'Laboratorio de Hemoderivados' were synthesized as a model of sustainable company, with an integral vision of social responsibility and sustainability, also considering that it is

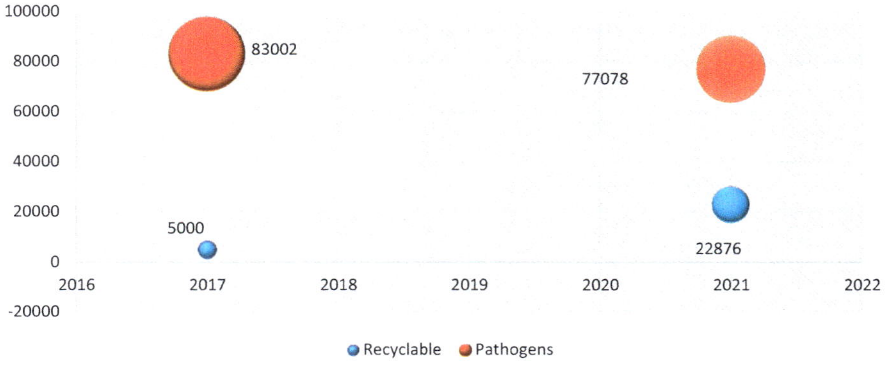

Fig. 5 Evolution of waste management (in kg) Source: Prepared by the authors

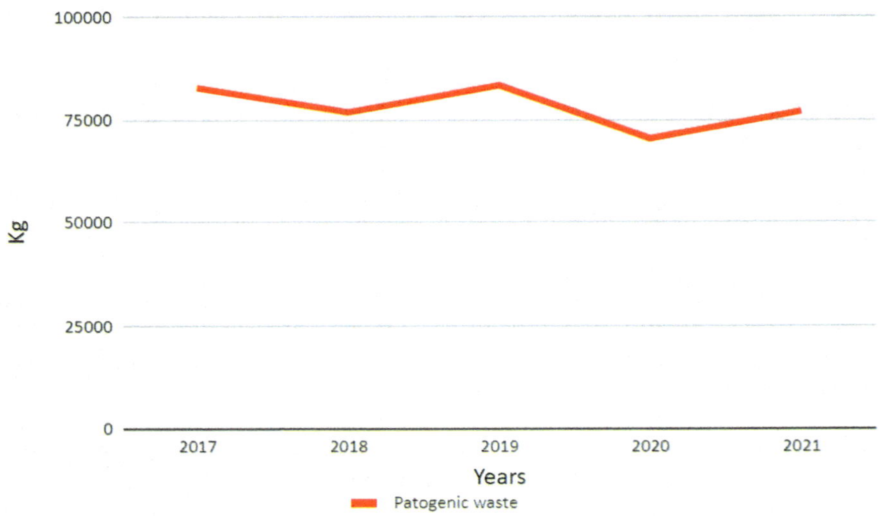

Fig. 6 Evolution of pathogenic waste Source: Prepared by the company

a public pharmaceutical company under university management. In order to improve its competitiveness and manage its business model within the framework of the SDGs, the laboratory established links with various stakeholders. The comparative analysis of the firm's sustainability reports allows us to measure the growing commitment assumed by the laboratory and, on the other hand, the case reports on various institutional processes that followed one another in the same direction and consolidated the commitments undertaken.

This synthetic and comparative presentation of the Laboratory's reports sought to account for the evolution and milestones of the change process, contributing to the understanding from the perspective of the actors and the idiosyncrasy of the pharmaceutical organization itself.

Thus, the analysis was aimed at the evolution of its business model, responsible consumption and production actions and sustainability reporting, from a systemic view that involved the various stakeholders related to the laboratory. All this with a focus on SDG 12 responsible production and consumption.

6 Discussion Questions

1. Identify the strategy/ies adopted by the LH to innovate towards a sustainable business model. Is there any archetype of those proposed by Ritala et al. (2018) predominant?

Table 1 SWOT analysis of Hemoderivados UNC according to the scope of sustainability

Strengths	Weaknesses	Challenges	Opportunities
Fast-growing sustainability policies in the organization	Low supplier development and sustainable logistics	Increased regulatory measures	Circular economy in development
Strong professional team qualified in sustainable management	Environment and sustainable culture in development	Pressure for cost reduction efforts linked to sustainable processes	Significant government commitment to sustainable management
Strong corporate governance support for sustainability policies	Characteristics of industrial processes (e.g. refrigerant gases/greenhouse effect)	Benchmarking key performance parameter	Increased information management system
High employee commitment to the organization	Improvement initiatives are voluntary	Demand for greater communication with stakeholders	Corporate governance committed to the issue
Internal environment favourable to the SDGs	Fear of non-return on investment in R&D and sustainability	Pressure for the adoption of new technology with a smaller environmental footprint	Availability of new technology for operational excellence
Government initiatives for strategic alliances	Low openness to data exchange	Meeting global regulatory demand	Process improvement for resource optimization and management simplification management

Source: Own elaboration

2. Identify and describe the actions taken by LH in relation to the fulfilment of SDGs 12 and 4 and their targets. Which ones do you consider most relevant? Justify your answer based on the SDGs.
3. Analyse the sustainability reports presented and establish a comparison in terms of the actions reported. Can you link the changes to any of the theoretical perspectives for incorporating sustainability?
4. What initiatives could LH take to improve its social, environmental and economic impact? What obstacles or constraints might the organization face in implementing these initiatives?
5. Assess the impact generated by the actions taken by LH and the importance of continuing to move forward in meeting new objectives.

Annex 1. SDG 4

There are actions that reflect the pharmaceutical industry's contribution to sustainable development. Perhaps the most relevant is the sustainable health care reflected in SDGs 3, 9 and 17. The goal is that every person who needs medicines should have access to them.

As an industrial sector with significant economic power, pharmaceuticals also contribute to the goal of responsible business reflected in SDGs 8, 16 and 17.

There are actions that reflect the pharmaceutical industry's contribution to sustainable development. The most relevant, perhaps, is the sustainable health care reflected in SDGs 3, 9 and 17. The goal is that every person who needs medicines should have access to them.

As an industrial sector with significant economic power, pharmaceuticals also contribute to the goal of responsible business reflected in SDGs 8, 16 and 17.

As a highly regulated sector, it must meet high standards of business ethics and integrity. Its global nature commits it to respect and engage with human rights, equality and gender issues and to work with rigorous codes of conduct. In this sense, there is concern for its employees and collaborators, reflected in SDGs 5 and 8. Promoting diversity and inclusion and a dynamic and developing work environment that encourages professional growth and opportunities, with a special focus on women, are among the most important contributions at the international level.

The permanent development of new drugs and clinical uses generates a link with universities and training institutes that extends to educational and ongoing training programs that contribute to quality education. The activity itself, given its characteristics, favours strategic alliances and the creation of jobs with actions and support for programs that improve and provide sustainable benefits to the communities in which it operates.

Concern for the environment, new energies and responsible consumption is reflected in SDGs 6, 7, 8, 9, 12 and 13. Minimizing the ecological footprint and increasing the use of renewable resources, while expanding the global growth of the pharmaceutical industry, is an imposition for the sector (Belkhir & Elmeligi, 2019; Egan, 2018; Santos Bravo et al., 2013).

Trend of Practices Related to Sustainability

Goal 3 Health and Wellness: The pharmaceutical industry plays a leading role, to the point that almost all available medicines are the product of its R&D and that new medicines are responsible for 73% of the prolongation of life expectancy achieved in recent decades. Between 2000 and 2009, life expectancy in OECD countries increased by 1.74 years, with 1.27 years being the result of pharmaceutical innovation. With regard to the goal of achieving universal health coverage (point 3.8 accessibility and availability of safe, effective and quality medicines and vaccines for all), the pharmaceutical industry is committed to expanding patients' access to treatment and, at the same time, achieving the long-term sustainability of the health sector.

Goal 4 Ensure inclusive, equitable and quality education: The pharmaceutical industry is a leader in R&D investments and in the generation of skilled jobs with demand for university graduates. Scientific research and the improvement of technological capacity, all of which are hallmarks of the pharmaceutical industry, require the development of alliances with university researchers in projects with innovative potential for pharmaceutical companies; in addition, lifelong learning and the generation of knowledge mobilize the sector to obtain quality education from a sustainable development perspective (target 4.7).

Chronology of Significant Initiatives in Sustainability

In 2017, within the framework of SDG 4 (Fig. 7), in addition to promoting adherence to the Global Compact, the LH organized the second Latin American Meeting of Plasma and Hemoderivatives, the Primary Immunodeficiencies Conference, aimed at professionals, patients and relatives, and the First Meeting of Plasma Donors and Patients.

It also participated in congresses, conferences and specialized scientific journals, with scientific papers and dissertations. It also participated in conferences of professional associations, such as the Asthma, Allergy and Immunology Association of Córdoba (AAAeIC).

In 2018, meanwhile, the meeting of patients and families for the World Haemophilia Day, the IV Theoretical and Practical Workshop on Hemophilia and

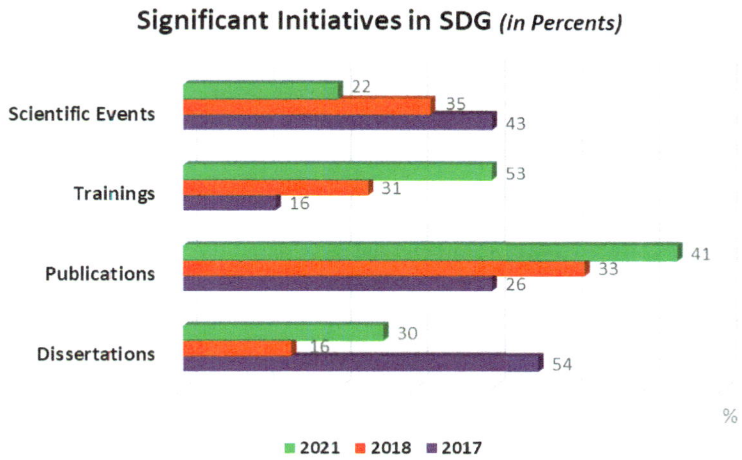

Fig. 7 SDG 4 main initiatives Source: own elaboration

its treatments, aimed at professionals, patients and families, as well as the update workshops on hemoderivative medicines: clinical cases and pharmacovigilance, and the one aimed at traumatologists stand out.

In addition to participating in congresses, conferences and specialized scientific journals, with scientific papers and dissertations, this year the plasma donation campaign was also disseminated at the TEDx Córdoba talks (artistic intervention) and at the Regional Conference on Higher Education (information stand).

In the triennium 2019–2021, meanwhile, activities continued with patients, family members and professionals in the framework of conferences and training workshops in pharmacoepidemiology, as well as in the awareness and identification of haemophilia, and with the scientific conference 'First regional preparation of Subcutaneous Gamma Globulin', in the framework of the launch of this new product, which also included a workshop on use, pharmacovigilance and prevention of risks associated with this drug.

The training of undergraduate and graduate students, professionals and personnel of the LH was materialized in this period in the start-up of the UNC Diploma in Industrial Production of Medicines; the organization of workshops in alliance with the Association of Patients with Primary Immunodeficiencies; and the continuity of the Practicanato for students of the undergraduate careers of the School of Chemical Sciences, including Biotechnology. Other importante iniciatives are the incorporation of interns and the implementation of a scholarship system for advanced students of the UNC.

Also included here are training courses on Pharmacovigilance; on the products of the tissue division; or those aimed at professionals of the University Maternity Hospital and the National Hospital of Clinics, for the procurement of amniotic membrane, in particular, and on the challenges in the procurement and ablation of organs and tissues, in general.

The scientific advances in the framework of the Public-Public Consortium between the Laboratory, the UNC and the Conicet (BioHemo), for the elaboration of proteins of therapeutic value from recombinants and genetic engineering technology, as well as the alliances with the Leloir Institute of ANLIS and the scientific-technological matrix of the UNC for the development and elaboration of a gamma globulin enriched with anti-Sars-CoV-2 antibodies, constitute another chapter in the framework of the contributions to SDG 4.

During this period, the LH also promoted and launched, together with the Secretariat of University Extension and the UNC Blood Bank, the training program for blood and plasma donation promoters 'Compartir es Salud' (Sharing is Health), which is being developed in different localities of the province of Córdoba. In addition, together with the Medicines Information Centre of the Faculty of Chemical Sciences (UNC), it prepared the digital publication 'Medicines: What they are, where they come from and how they are developed', aimed at the community in general.

Appendix: Teaching Notes for the Case

Synopsis

This case aims to contribute to the debate on the integration of sustainability and social responsibility (SR) in the management of organizations. This discussion has generally focused on the sustainability and SR actions carried out by companies (the what), or on the motivations they have for integrating them into management (the why), and less on understanding the change process faced by organizations to evolve towards 'sustainable business' models (the how).

This case proposes the approach of a sustainable business model, with an integral vision of social responsibility and sustainability (SR&S), from the perspective of a public pharmaceutical company under university management: *the Laboratorio de Hemoderivados of the Universidad Nacional de Córdoba-UNC (Argentina).*

The analysis is aimed at the evolution of its business model, responsible consumption and production actions and sustainability reports, from a systemic vision that involves the various stakeholders related to the laboratory, emphasizing SDG 12 responsible production and consumption based on its dual role as a public company and university.

Potential Audience and Instructor's Material

This case has been developed for use in the framework of marketing, CSR and business strategy courses in general and sustainable business model seminars in particular.

The potential audience for this chapter is twofold:

- Undergraduate students pursuing business administration or marketing studies
- Graduate students pursuing an MBA or Master's degree in CSR/sustainability or executive programs and seminars (on CSR in general and sustainable business model/business and SDGs in particular).

Learning Objectives

1. To contribute to the understanding of responsible production as a relevant dimension of the sustainability policy of an industrial organization in the pharmaceutical sector.
2. To describe the approach to SDG 12 (Responsible production and consumption) by an industrial organization inserted in a university institution.
3. To analyse valuable initiatives of the organization to contribute to the sustainable development and human quality of the community.
4. To assess the evolution and impact of the sustainable development actions carried out by the organization.

Time Frame for Class Discussion

This case can be delivered in a 60–90-min session, depending on the number of students participating in the discussion activities as well as the previous work students have done prior to the class.

Suggested Discussion Questions

1. Identify the strategy/s adopted by the LH to innovate towards a sustainable business model. Is there any archetype of those proposed by Ritala (2018) predominant?
2. Identify and describe the actions taken by LH in relation to the fulfilment of SDG 12 and SDG 4 and their targets. Which ones do you consider most relevant? Justify your answer based on the SDGs.
3. Analyse the sustainability reports presented, and establish a comparison in terms of the actions reported. Can you link the changes to any of the theoretical perspectives for incorporating sustainability?
4. What initiatives could LH take to improve its social, environmental and economic impact? What obstacles or constraints might the organization face in implementing these initiatives?
5. Assess the impact generated by the actions taken by LH and the importance of continuing to move forward in meeting new objectives.

Analysis

Answer to Question 1. *Identify the strategy/s adopted by the LH to innovate towards a sustainable business model. Is any of the archetypes* (Ritala et al. 2018) *predominant*?

The student could elaborate on the archetypes defined by Ritala et al. (2018). There is not one dominant one, but several are worked on simultaneously.

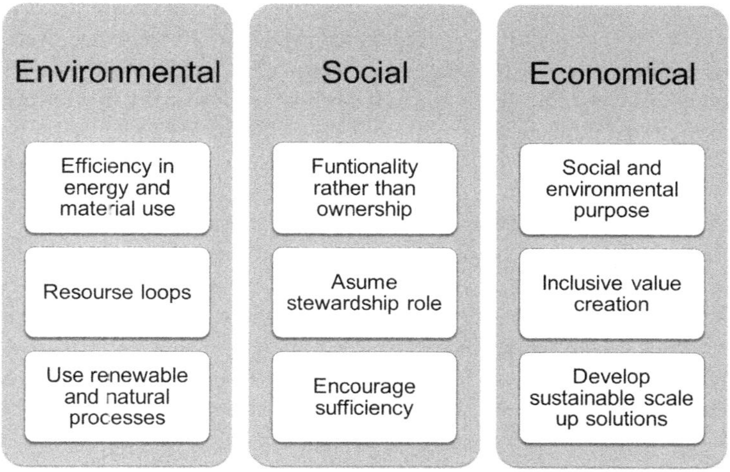

Source: adapted from Ritala et al. (2018)

At the environmental level, we are mainly working to improve energy efficiency and close the loop through circular economy actions.

At the social level is the adoption of stewardship role or surveillance, from the promotion of the population, and the promotion of sufficiency from consumer education with plasma donation campaigns.

Finally, at the economic level, repurpose for society is a non-profit public entity while developing scalable solutions, from a business model based on the development of partnerships with various health institutions throughout the country.

Answer to Question 2. *Identify and describe the actions carried out by LH in relation to the fulfilment of SDGs 12 and 4 and their targets. Which ones do you consider most relevant? Justify your answer based on the SDGs.*

At this point, the student could point out the statement of both SDGs 12 and 4 and their targets, linking to the information provided in the case.

With respect to SDG 12, they could point out the various actions linked to responsible production and circular economy.

The 2030 Agenda, in its SDG 4, recognizes quality education as a means to achieve the rest of the SDGs, with education for sustainable development as a fundamental tool in target 4.7. Thus, for example, with respect to this target, you can highlight the contribution of LH with different training actions.

This is a free-response question (no right or wrong answers). Each student can give his or her personal opinion. The most important aspects of this question are linked to facilitate the student's argumentation, either favourable or unfavourable to the LH actions, and that they reflect on what may be the strengths and weaknesses (or pros and cons).

Some ideas for working on this question might be the following: probe for the link between SDGs and universities. The SDGs cover a broad set of social, economic and environmental challenges, which are closely related, and while governments and businesses have responded to them, the expertise of universities is crucial to their achievement. According to this network, the SDGs will not be achieved without these institutions and consider that universities must 'make visible and maintain public commitment to the Agenda, being an intersectoral facilitator for the mobilization, design and implementation of policies' (Miñano & Garcia, 2020, p. 6).

Answer to Question 3. *Analyse the sustainability reports presented with respect to the reported actions. Can you link the changes to any of the theoretical perspectives for incorporating sustainability?*

This is a free-response question (no right or wrong answers). Each student can give his or her personal opinion as soon as he or she substantiates it with the theoretical axes. In terms of the actions reported, there is a growth of these actions and an alignment with the SDGs.

The student could rely on the different theoretical perspectives to incorporate sustainability in organizations. Public organizations are usually called upon to be accountable for their actions and to meet society's expectations, so they could be justified from the theory of legitimacy and stakeholders.

The Theory of Legitimacy is linked to the company's need to be authorized to operate (Deegan, 2002), which may be affected if it does not operate in an acceptable manner. This acceptability, in turn, is linked to the thinking of stakeholders, which is why Stakeholder Theory, or the mention of stakeholders, has been the starting point of numerous publications. Stakeholder Theory is based precisely on taking into account the perspectives and expectations of stakeholders (Freeman et al., 2007). Among the expectations is that of receiving information from companies, not only financially but also about their sustainable business conduct (Hahn & Kühnen, 2013). Thus, by providing information linked to the sustainability actions it performs, the company can shape perceived legitimacy (Campbell et al., 2003), an action that Hahn and Kühnen link to Signalling Theory (2013).

As for the inclusion of the SDG framework within the report, it could be justified from the Signalling Theory. Signalling Theory suggests that when there is asymmetric distribution of information, one party tries to convey credible information about itself to the other party (Connelly et al., 2011). Signalling—for example, its social and environmental commitment to stakeholders—would give companies a secure competitive position and a possible competitive advantage (Nunes & Park, 2017).

Answer to Question 4. *What initiatives could LH take to improve its social, environmental and economic impact? What obstacles or constraints might the organization face in implementing these initiatives?*

This is a free-response question (no right or wrong answers). Each student can give his or her personal opinion. The most important aspects of this question are linked to students strongly arguing their opinions, whether favourable or unfavourable to LH's actions, and reflecting on the obstacles or constraints it might face.

Some ideas for working this question might include the following:

The transformation of consumption patterns, as well as of productive systems, has become the most relevant change movement for the achievement of sustainable development and the goals of the 2030 Agenda (Yildirim, 2022). This movement is associated with booming concepts such as responsible consumption, sustainable living, education for sustainable development and sustainable business models.

In terms of restrictions, issues related to monetary resources, capacities, etc., could be highlighted.

Answer to Question 5. *Evaluate the impact generated by the actions carried out by LH in relation to responsible consumption and the importance of continuing to advance in the fulfilment of new objectives.*

This is a free-response question (there are no right or wrong answers). Each student can give his or her personal opinion. The most important aspects of this question are to address the concept of responsible consumption and possible actions related to it carried out by the laboratory.

Some ideas for working on this question could be the following:

Responsible consumption is defined as 'consumption that has less negative or more positive impact on the environment, society, the individual and other beings' (Ulusoy, 2016), being an 'umbrella' concept that encompassed several specific consumption practices, including sustainable consumption and consumer citizenship among others. Phipps et al. (2013) defined sustainable consumption as that which simultaneously optimizes the environmental, social and economic consequences of its purchases, use and disposal considering future generations.

Schrader (2007) addressed the preconditions of consumers' moral responsibility for sustainable consumption, starting from the concept of consumer citizenship. According to CNN (Thoresen, 2005, p. 7), this notion is linked to the inclusion of ethical, social, economic and ecological considerations in decision-making, 'the consumer citizen actively contributes to the maintenance of fair and sustainable development by caring and acting responsibly at the family, national and global level'. The concept of sustainable living first appears in the Sustainable Development Goals (SDG 4 Education and SDG 12.8 Responsible Consumption), it is linked to understanding how our lifestyle choices impact the world around us and also to finding ways for us all to live better and lighter lives. The call to apply the 'people lens' to sustainability is new and represents great opportunities for organizations called upon to contribute to sustainable development (UNEP, 2022).

References

Agarwal, R., Mittal, N., Patterson, E., & Giorcelli, M. (2021). Evolution of the Indian LPG industry: Exploring conditions for public sector business model innovation. *Research Policy, 50*(4), 104196.

Belkhir, L., & Elmeligi, A., (2019). Carbon footprint of the global pharmaceutical industry and relative impact of its major players. *Journal of Cleaner Production, 214*. https://doi.org/10.1016/j.jclepro.2018.11.204.

Campbell, D., Craven, B., & Shrives, P. (2003). Voluntary social reporting in three FTSE sectors: a comment on perception and legitimacy. *Accounting, Auditing & Accountability Journal, 16*(4), 558–581.

Connelly, B. L., Certo, S. T., Ireland, R. D., & Reutzel, C. R. (2011). Signaling theory: A review and assessment. *Journal of Management, 37*(1), 39–67.

Crane, A., & Matten, D. (2004). *Business ethics. A European perspective: Managing corporate citizenship and sustainability in the age of globalization*. Oxford University Press.

Deegan, C. (2002). Introduction: The legitimising effect of social and environmental disclosures – a theoretical foundation. *Accounting, Auditing & Accountability Journal, 15*(3), 282–311. https://doi.org/10.1108/09513570210435852

Egan, T. (2018). *Achieving Sustainability in an Evolving Pharma Sector: Life sciences companies look to utility savings, collaborative product development and serialization to go green*. Pharmamanufacturing – articles. https://www.pharmamanufacturing.com/articles/2018/achieving-sustainability-in-an-evolving-pharma-sector/.

Ferguson, T., & Roofe, C. G. (2020). SDG 4 in higher education: Challenges and opportunities. *International Journal of Sustainability in Higher Education, 21*(5), 959–975.

Franco, I., Saito, O., Vaughter, P., Whereat, J., Kanie, N., & Takemoto, K. (2019). Higher education for sustainable development: Actioning the global goals in policy, curriculum and practice. *Sustainability Science, 14*(6), 1621–1642.

Freeman, R. E., Martin, K., & Parmar, B. (2007). Stakeholder capitalism. *Journal of Business Ethics, 74*(4), 303–314.

Glavič, P. (2021). Evolution and current challenges of sustainable consumption and production. *Sustainability, 13*(16), 9379. https://doi.org/10.3390/su13169379

Hahn, R., & Kühnen, M. (2013). Determinants of sustainability reporting: A review of results, trends, theory, and opportunities in an expanding field of research. *Journal of Cleaner Production, 59*, 5–21.

Hartley, J. (2006). *Innovation and its contribution to improvement: A review for policymakers, policy advisers, managers and researchers*. Department for Communities and Local Government.

Izquierdo Maldonado, C., Vaca Aguirre, I. P., & Mena Campar, R. E. (2019). El nuevo sujeto social del consumo responsable. *Estudios De La Gestión: Revista Internacional de Administración, 4*, 97–123. https://doi.org/10.32719/25506641.2018.4.5

Milanesi, M., Andrea, R. A., & Guercini, S. (2020). Pharmaceutical industry riding the wave of sustainability: Review and opportunities for future research. *Journal of Cleaner Production, 261*, 2020.

Miñano, R. & García Haro, M. (Editores). (2020). *Implementando la Agenda 2030 en la universidad. Casos inspiradores*. Madrid: Red Española para el Desarrollo Sostenible (REDS), 2020.

Neto, B. (2012). LCA tool for sustainability evaluations in the pharmaceutical industry. *Chemical Engineering Transactions*.

Nunes, M. F., & Park, C. L. (2017). Self-claimed sustainability: Building social and environmental reputations with words. *Sustainable Production and Consumption, 11*, 46–57.

Osterwalder, A., & Pigneur, Y. (2010). *Business model generation*. Wiley.

Pan, L., Xu, Z., & Skare, M. (2022). Sustainable business model innovation literature: A bibliometrics analysis. *Review of Managerial Science*. https://doi.org/10.1007/s11846-022-00548-2.

Pérez Espinoza, J. M., Espinoza Carrión, C., & Peralta Mocha, B. (2016). La responsabilidad social empresarial y su enfoque ambiental: Una visión sostenible a futuro. *Revista Universidad y Sociedad [seriada en línea], 8*(3), 169–178. http://rus.ucf.edu.cu/

Phipps, M., Ozanne, L. K., Luchs, M. G., Subrahmanyan, S., Kapitan, S., Catlin, J. R., et al. (2013). Understanding the inherent complexity of sustainable consumption: A social cognitive framework. *Journal of Business Research, 66*(8), 1227–1234.

Rajagopal Ramachandran. (2017). Sustainability in the pharmaceuticals and fine chemicals industry, In: Martin A. Abraham (Ed.), *Encyclopedia of sustainable technologies*. Elsevier, pp. 675–681, ISBN 9780128047927, https://doi.org/10.1016/B978-0-12-409548-9.10255-6.

Ritala, P., Huotari, P., Bocken, N., Albareda, L., & Puumalainen, K. (2018). Sustainable business model adoption among S&P 500 firms: A longitudinal content analysis study. *Journal of Cleaner Production, 170*, 216–226.

Santos Bravo, S., Margarida, A., & Crespo de Carvalho, J. (2013). Understanding pharmaceutical sustainable supply chains – a case study application. *Independent Journal of Management & Production, 4*(1), enero-junio, 2013, pp. 228–247.

Schrader, U. (2007). The moral responsibility of consumers as citizens. *International Journal of Innovation and Sustainable Development, 2*(1), 79–96.

Thoresen, V. W. (ed.) (2005). Consumer Citizenship Education – Guidelines, Vol. 1. Higher Education, CCN (The Consumer Citizenship Network) (2005) Hamar, Norway.

Ulusoy, E. (2016). Experiential responsible consumption. *Journal of Business Research, 69*(1), 284–297.

Veleva, V., Hart, M., Greiner, T., & Crumbley, C. (2003). Indicators for measuring environmental sustainability: A case study of the pharmaceutical industry. *Benchmarking: An International Journal, 10*, 107–119. https://doi.org/10.1108/14635770310469644

Vilariño, M. V., & Bigorito, S. A. (2014). *Visión 2050 Argentina: Una nueva agenda de desarrollo*. Buenos Aires. Consejo Empresario Argentino para el Desarrollo Sostenible CEADS, 2014. http://www.ceads.org.ar/wp-content/uploads/2020/10/Vision-2050-Argentina.-Una-nueva-agenda-de-desarrollo.pdf

Weadon, M. (2010). Sustainability — An opportunity for the pharma industry. *Pharmaceutical Online Magazine*. https://www.pharmaceuticalonline.com/doc/sustainability-an-opportunity-for-the-0001

Yildirim, S. (2022). The consumer role for sustainable development: How consumers contribute sustainable development goals. In *Research anthology on measuring and achieving sustainable development goals* (pp. 872–888). IGI Global.

Fair Trade: A Path Towards Responsible Consumption and Sustainability. The Case of Medicus Mundi

Pilar Fernandez-Ferrin, M. Mercedes Galan-Ladero, Sandra Castro-Gonzalez, and Belén Bande

Learning Objectives

1. To know what Fair Trade is and how it can contribute to responsible production and consumption, and sustainability
2. To study the situation of Fair Trade in Spain
3. To analyze the NGDO Medicus Mundi and, mainly, the activities related to Fair Trade that it develops

1 Introduction

1.1 What Is Fair Trade? Concept and Principles

Fair Trade (FT) is a type of trade alternative to conventional one, which arose out of concern about the inequality generated by international trade and its negative consequences[1]: benefits were not distributed fairly, labor conditions and wages did not improve at the same rate as prices, a small number of companies dominated

[1] For more information, see the publication on Fair Trade by Fernandez-Ferrin et al. (2021), available at https://comerciojusto.org/publicacion/el-comercio-justo-en-euskadi-comportamiento-y-motivaciones-de-compra/ Or, directly at https://comerciojusto.org/wp-content/uploads/2022/03/investigacion-cj-en-euskadi-def.pdf

P. Fernandez-Ferrin
University of the Basque Country UPV/EHU, Vitoria, Spain
e-mail: pilar.fernandezf@ehu.eus

M. M. Galan-Ladero (✉)
University of Extremadura / Universidad de Extremadura, Badajoz, Spain
e-mail: mgalan@unex.es

S. Castro-Gonzalez · B. Bande
Universidade de Santiago de Compostela, Lugo, Spain
e-mail: sandra.castro@usc.es; belen.bande@usc.es

E. C. Bianchi et al. (eds.), *Responsible Consumption and Sustainability*, Springer Business Cases, https://doi.org/10.1007/978-3-031-30742-3_17

Fig. 1 10 Principles of Fair Trade. Source: WFTO (2022)

1.- Opportunities for Disadvantaged Producers.
2.- Transparency & Accountability.
3.- Fair Trade Practices.
4.- Fair Payment.
5.- No Child Labor. No Forced Labor.
6.- No Discrimination. Gender Equity. Freedom of Association.
7.- Good Working Conditions.
8.- Capacity Building.
9.- Promote Fair Trade.
10.- Climate Action and Protection of the Environment.

international markets and had the power to set commercial conditions and prices, and there were unfavorable and harmful impacts on the environment (resulting from massive and intensive production). Thus, FT is often associated "*with small producers, with a 'fair' price to be received by these people, as well as with organic, traditional, and sustainable production*" (Fernandez-Ferrin et al., 2021: 9).

Therefore, Fair Trade is defined as "*a trading partnership, based on dialogue, transparency, and respect, that seeks greater equity in international trade. It contributes to sustainable development by offering better trading conditions to and securing the rights of marginalized producers and workers—especially in the South*" (WFTO, 2022). Consequently, the generic objectives of FT are as follows (WFTO, 2022): inclusive economic growth, decent work, "people first" trade policies, gender equity, food security, sustainable livelihoods, ecological balance, and thriving communities.

In short, FT is a global movement based on production and marketing methods that prioritize the needs of people and the planet rather than strictly economic benefits. It facilitates producers' social and economic development, strengthening self-management and connecting them with consumers through supply chains that are as short and transparent as possible, promoting women's empowerment, adopting circular economy production processes, and supporting sustainable and organic agriculture (WFTO, 2022). All of this is embodied in the 10 FT Principles (Fig. 1).

Furthermore, FT is often considered synonymous with terms such as alternative, solidarity, ethical, or responsible trade. Although all these concepts emerged as an alternative to traditional, industrial, or mass trade (to alleviate its negative effects on society and the environment) and are interconnected, there are differences between them (Fernandez-Ferrin et al., 2021). Thus, FT is only one concrete form of alternative and social/solidarity-based trade and one of the possible manifestations of responsible consumption. Furthermore, it is important to differentiate FT products from other alternatives, such as biological or organic products. While the latter comply with one of the 10 FT Principles, namely, respect for the environment and biodiversity and the promotion of eco-sustainable and organic production, they may or may not comply with the rest of the principles. On the other hand, not all Fair Trade-certified products are organic, although there is an upward trend.

1.2 Origin and Evolution of Fair Trade: New Trends

The first FT initiatives took place in the United States in the 1940s and 1950s, although it was not until 1958 that the first FT store was opened in that country (WFTO, 2022). At the same time (in the late 1950s), FT took its first steps in Europe.

In the following decades, the 1970s and 1980s, FT experienced great growth, developing and being introduced in more countries, favoring the emergence of many producer organizations and communities.[2] In the decade of the 1990s, the consolidation of FT took place. Since then and until today, FT has become, in addition to solidarity and alternative trading system, a global movement (CECJ, 2022), with an impact (direct and/or indirect) on practically all the 17 SDGs: two and a half million small-scale producers and workers; organized in more than 4,000 organizations, in more than 70 countries (WFTO, 2022); and whose products are sold in thousands of stores specialized in FT products but also supermarkets and hypermarkets, and other points of sale (fairs, flea markets, etc.), as well as on the Internet (online stores); and with growing participation in international forums and spaces for debate (Table 1).

Although FT initially focused on handicraft products, food has subsequently become the main FT products, with coffee and chocolate being the "star" products (Fernandez-Ferrin et al., 2021). Currently, cosmetics and personal hygiene products, toys, cleaning products, books, and stationery are also marketed within FT.

Although FT has traditionally been associated mainly with small producers in the South, new trends in FT include Domestic Fair Trade (or Local Fair Trade), aimed at strengthening the position of small local producers who have been marginalized by the mainstream market, as a means to preserve the culture of agriculture and the rural community, promote economic democracy and diversity, and ensure a healthier and more sustainable planet.

1.3 Fair Trade in Spain

It was not until the 1980s that the first FT initiatives appeared in Spain. However, in the following decades, it has been consolidated (Table 2), experiencing continuous and significant growth (although it can still develop much more and reach a wider public).

Some of the most relevant characteristics of FT in Spain are the following (Fernandez-Ferrin et al., 2021):

1. Participating agents. These include producers, importers (cooperatives, NGOs, companies) and traders, FT stores, consumers, and FT organizations and networks.

---

[2]For more information, see Coordinadora Estatal de Comercio Justo—CECJ (2022). Available at: https://comerciojusto.org/

Table 1 Chronology of key events related to Fair Trade

Decades	Events
1940s	First FT initiatives in the United States
1950s	1958—Opening of the first FT store in the United States
	First steps of FT in Europe
1960s	More FT initiatives are developed
1970s	1973—FT coffee is introduced
1980s	1987—The European Fair Trade Association (EFTA) is constituted
	1988—First FT guarantee seal in the Netherlands (Max Havelaar Seal)
	1989—International Federation of Alternative Trade (IFAT) is created
1990s	1994—The Network of European World Shops (News!) is constituted
	1997—Fairtrade Labelling Organizations International (FLO) is created Objective: To establish global standards and certification processes for FT products and examine and certify their production and marketing
2000s	2002—FLO launched a new international certification mark for certain FT products (International Fairtrade Certification Mark) to unify the different labels used by its member organizations, avoid confusion among consumers, and facilitate their international marketing
	2003—IFAT changes its initial name but keeps its acronym (it is now called the International Fair Trade Association)
	2004—IFAT launches the first certification seal for Fair Trade organizations
	2008—IFAT is renamed the World Fair Trade Organization (WFTO)
	2009—WFTO and FLO jointly publish the International Fair Trade Charter
2010s	2011—FLO changes its name to Fairtrade International
	2013—New WFTO Guarantee System, to evaluate the organization as a whole (according to the 10 Fair Trade Principles)
	2016—Launch of WFTO Product Label (seal for FT products from guaranteed FT organizations)
2020s	Fair Trade: Key to progress toward the SDGs

Note: Evolution of the names of the main Fair Trade organizations:
International Federation of Alternative Trade (IFAT)—International Fair Trade Association (IFAT)—World Fair Trade Organization (WFTO)[a]
Fairtrade Labelling Organizations International (FLO)—Fairtrade International[b]
Source: Own elaboration, based on Fernandez-Ferrin et al. (2021)
[a]WFTO is the global network of producers, marketers, exporters, importers, wholesalers, and retailers that demonstrate 100% commitment to Fair Trade and apply the 10 WFTO Principles of Fair Trade to their supply chain. Nowadays, WFTO operates in over 70 countries across five regions (Africa, Asia, Europe, Latin America, North America, and the Pacific Rim) with elected global and regional boards (WFTO, 2022). For more information: https://wfto.com/
[b]Nowadays, Fairtrade International is the organization that coordinates the Fairtrade labeling scheme (Fairtrade International, 2022). It is co-owned by three regional producer networks (representing over 1400 organizations certified to Fair Trade standards in 73 countries) and 19 national Fair Trade organizations that license the Fair Trade label and promote it to businesses and consumers (WFTO, 2022). For more information: https://www.fairtrade.net/

Table 2 Chronology of key events related to Fair Trade in Spain

Decades	Events
1980s	First FT initiatives in Spain
1990s	Most organizations that will make up the FT movement in Spain are formed It begins to carry out an important work of awareness and disseminating FT
	1995—The Network of Networks of Alternative and Solidarity Economy (REAS) was born in Spain
	1996—The State Coordinator of Fair Trade (*Coordinadora Estatal de Comercio Justo*) in Spain is created
2000s	SETEM[a] carried out the first sales analysis: 7 million euros in turnover, almost 70% of purchases were made in FT stores, and more than half corresponded to handicraft items Oxfam Intermon sells the first pallet of coffee to a large distribution chain, thus initiating the expansion to other sales spaces
	2005—Fairtrade Spain (today, Fairtrade Iberica, which also includes Portugal), the Fairtrade label association integrated into Fairtrade International, is founded
	2008—The expansion of certified products in companies and large retail and hospitality chains is taking off
2010s	Continued and significant growth of FT sales in Spain
	2019—Sales exceeded 100 million euros in the country (138.7 million, exactly)
2020s	2020–2021—The COVID-19 pandemic slowed down the rapid growth of FT sales in Spain up to that point, although at least the figures are maintained 2022-The "More Fair Trade, More SDGs" campaign is launched.

Source: Own elaboration, based on Fernandez-Ferrin et al. (2021) and CECJ (2022)
[a]SETEM is a federation of international solidarity NGOs that seeks to raise awareness of North-South inequalities, denounce their causes, and promote social transformations, both individual and collective, to achieve a fairer and more united world. For more information, see: https://www.setem.org

2. Sales growth. Sales of FT products have been growing significantly and continuously over the last decades,[3] although in recent years, this growth has slowed down.[4] In addition, there are still barriers to the purchase of these products.[5]
3. Diversification/expansion of points of sale. Traditionally, FT specialty stores have been the preferred place for consumers to buy these products. However, in recent years, they have lost their representativeness (only just over 3% of sales,

[3]Since 2000, when FT data collection began, sales have increased 14-fold (Fernandez-Ferrin et al., 2021).

[4]Thus, for example, for the period 2017–2022, sales increased from 51.3 million euros in 2017 to 78.1 million euros in 2018, 138.7 million euros in 2019, 143.7 million euros in 2020, and 144.1 million euros in 2021. On the other hand, the average consumption per inhabitant per year of Fair Trade products with Fair Trade labels in Spain was, in 2021, 2.90 euros (far from other European countries—and even Portugal—which is integrated into Fairtrade Iberica and has an average of 4.53 euros per inhabitant per year).

[5]These barriers include the lack of clear labeling of these products, the lack of information and knowledge of FT products and producers, and an allegedly high price (Fernandez-Ferrin et al., 2021: 10).

Table 3 Main activities

Awareness-raising initiatives, both in person and in virtual format (due to the pandemic and to reach new audiences): These include conferences and debates (mainly online), exhibitions and gastronomic routes, cooking workshops, chocolate tastings, natural cosmetic workshops, reading clubs in libraries, and cultural activities (theater, concerts, etc.,—although limited due to the pandemic).
Work in educational centers at all levels (primary, secondary, high school, vocational training, university, and others).
Communication and dissemination of FT, through websites, profiles in social networks (campaigns in social networks have been promoted), and virtual initiatives.
Generation of knowledge (reports and publications).
Denouncing situations of violation of human and labor rights by companies.
Political advocacy (to influence decisions within political, economic, and social institutions).
Networking (to interact with other organizations).
Training of volunteers.

Source: Own elaboration, based on CECJ (2022)

although worth 4.71 million euros in 2021), and supermarkets and department stores have become the preferred channel for the distribution of FT products (79.4% of sales in 2021, worth 7 million euros). The HORECA channel (hotels, catering, and vending machines) is also used, with 13.1% of sales in 2021, small retailers (organic stores, herbalists, conventional establishments), as well as fairs and flea markets, and online stores, which are already common outlets for these products. Direct sales are also made to companies or public administrations (1.4% of sales in 2021)—CECJ (2022).

4. Profile of FT product buyers. They tend to be people with a higher level of education, who show a greater knowledge of these products, have a clearer perception of FT products, and give greater importance to the values they represent. They perceive fewer barriers to purchase, with price and assortment playing a relative role, and show a higher intention to purchase in the future.
5. Main activities carried out. They aimed at promoting among citizens another way of consuming, raising awareness about the objectives of FT. Also, they demand political and legal measures that prioritize the rights of people and the protection of the planet over economic benefits (CECJ, 2022). Thus, for example, social awareness and mobilization initiatives, dissemination, denunciation, and political advocacy stand out (CECJ, 2022)—Table 3.
6. Types of products: The main products are food (96.7% of turnover in 2021), with cocoa (and sweets) being the "star" product, followed by coffee and tea. Handicrafts, which include textiles (clothing and home), accessories, and decoration, bijoux/jewelry, and toys, accounted for 3% of turnover in 2021. Others, including cosmetics, accounted for only 0.3% (CECJ, 2022).

2 Case Development: Medicus Mundi and Fair Trade

2.1 History[6]

Medicus Mundi International (MMI)—"Health for All" is a network of organizations[7] working in the field of international cooperation in health.[8] The network promotes access to health and health care as a fundamental human right and is a nonstate actor in official relations with the World Health Organization (WHO).

Promoting "Health for All" implies (1) promoting universal health coverage, (2) strengthening public health systems, (3) favoring health policies that include social determinants and evidence-based health practices, and (4) calling for global governance for health.

Its main objective, therefore, is "to eradicate poverty so that everyone can live in dignity and enjoy the right to health." To achieve this goal, Medicus Mundi accompanies communities by implementing projects that serve to improve their health[9] but also contribute to the construction of global citizenship committed to social change, fostering a culture of solidarity and citizen commitment. Table 4 shows its main values.

Currently, it has three main lines of work: (1) development and humanitarian aid projects: health projects, to promote local capacities, and development interventions, to promote human rights; (2) Education for social transformation: to promote active and committed citizenship in the struggle for social justice; and (3) Fair Trade: a commitment to quality products and an economic model that ensures decent working conditions and remuneration.

2.2 Federation of Medicus Mundi Spain

The Federation of Medicus Mundi Spain is a nonprofit organization of international solidarity with an internal federal structure (it is made up of independent associations

[6]This section is based on information provided by Medicus Mundi on its official websites: https://www.medicusmundi.es and https://www.medicusmundi.org (last access: December 30, 2022).

[7]19 members, in 2019 (Medicus Mundi, 2022).

[8]Based in 12 countries (Austria, Belgium, Benin, Germany, Italy, Kenya, the Netherlands, Poland, Spain, Switzerland, the United Kingdom, and Zimbabwe)—Medicus Mundi (2022).

[9]In its broadest sense, health depends, among other aspects, on having access to clean drinking water; having adequate sanitary conditions; enjoying healthy food, adequate nutrition, and housing; having decent conditions at work and in the environment; and being able to access education (including that related to health). For this reason, Medicus Mundi points out that it focuses on eliminating avoidable differences related to social, economic, gender, and environmental, governmental, and community variables. To achieve this, its main commitment is to strengthen public health systems.

Table 4 Medicus Mundi's main values

Values	Description
Participation	To build equal partnership relationships based on trust and mutual respect
Independence	Not to depend on any political, economic, or religious group
Integrity	Shape credibility based on honesty and transparency
Professionalism	Seriousness and rigor in commitments and actions, providing an effective, efficient, and quality response to the needs of the most disadvantaged populations
Plurality	Relationships are based on respect for different political, religious, and philosophical positions because it favors the search for alternatives
Solidarity	To create a more just and responsible world, achieving a social transformation that improves the living conditions of the most impoverished societies and fully exercising the rights of individuals and peoples

Source: Medicus Mundi (2022)

that act with autonomy but share a common vision and proposals). In 2021, the association had 4,034 members, 366 volunteers, 76 employees, 662 donors, 83 counterparts, 63 public funders, and 22 private entities and/or groups (Medicus Mundi, 2022). According to its 2021 report, Medicus Mundi worked that year on a total of 149 projects, developed in 19 countries in Latin America, Africa, Asia, and Eastern Europe, with a total managed budget of 13,060,075 euros and reaching more than 1,540,453 people (of whom 941,858 were women). The main events in its history are listed in Table 5.

2.3 Medicus Mundi and Fair Trade

Medicus Mundi belongs to the State Coordinator of Fair Trade ("Coordinadora Estatal de Comercio Justo," in Spanish) and is also part of the World Fair Trade Organization (WFTO). It believes that citizens, as consumers, have power and can contribute to social transformation. Thus, through their daily consumption habits and through Fair Trade, they can achieve it, contributing to a fairer and more supportive alternative economic model, which ensures decent working conditions and remuneration to producers, that child labor has not been used, that its production process is respectful with the environment, and that there is gender equality when it comes to receiving the salary for the same work.

In this line of Fair Trade, Medicus Mundi Spain has developed different awareness campaigns. It has an annual awareness-raising program, which it carries out on several fronts, but there are also specific Christmas campaigns (as this is the time of year when they sell the most). Some of these campaigns are carried out in collaboration with the State Coordinator of Fair Trade and other NGOs. Table 6 shows some of the most relevant initiatives carried out in recent years in one of its territorial associations (Medicus Mundi Alava).

Table 5 The main events in Medicus Mundi Spain's history

Decades	Events
1960s	Medicus Mundi Spain, the Spanish branch of Medicus Mundi International, was created in 1963. It is the first Spanish NGDO specialized in health
	Growth of the association: Branches are opened in different Spanish cities, and interventions are carried out in different countries A Christian-inspired model of assistance work is imposed, sending volunteer medical personnel (usually for short stays), medical supplies, and medicines
1970s	The growth of the association continues
	Medicus Mundi was recognized by the World Health Organization (WHO) in 1974
	1978: The association's goal is "Health for all by the year 2000"
	1979: Medicus Mundi declares itself a professional, international, non-confessional, and non-governmental organization
1980s	1986: Medicus Mundi joined other NGOs and founded the Spanish Development NGO Coordinator (*Coordinadora de Organizaciones No Gubernamentales para el Desarrollo de España*—CONGDE; now, *La Coordinadora de Organizaciones para el Desarrollo*), thus betting on networking
1990s	It received the Prince of Asturias Award for Concord, together with Doctors Without Borders, in 1991
	The Federation of Medicus Mundi Associations in Spain was created in 1992
	Medicus Mundi begins to provide humanitarian aid (after the Rwandan genocide) in 1994
	Medicus Mundi was declared Public Utility in 1998
	Medicus Mundi was included in the Registry of Non-Governmental Development Organizations of the General Secretariat of the Spanish Agency of International Cooperation for Development (*Agencia Española de Cooperación Internacional para el Desarrollo*—AECID) in 1999
2000s	2002: Medicus Mundi proposes integrating NGDOs into public health systems at all levels (central, regional, and local)
	2003: Medicus Mundi developed its first strategic plan, which incorporates the concept of sustainable human development and health advocacy as a fundamental element in defense of the right to health of all people It also begins to prepare studies and reports
	It received accreditation as a qualified NGDO from AECID in 2005
2010s	2013: Medicus Mundi celebrates its 50th anniversary, with more than 1,000 health cooperation projects in 60 countries, 100 health personnel training programs, and more than 150 awareness-raising, development education and advocacy campaigns
	A process of merging associations was initiated in 2015, in order to join efforts (due to the cuts in the development cooperation policy caused by the economic crisis that had started in 2008)
	2019: The headquarters in Alava also opens its online FT store, with shipments throughout the peninsula
2020s	2022: The Federation of Medicus Mundi in Spain comprises 8 territorial associations that, although they act autonomously, share a common ideology and purposes.

Source: Medicus Mundi (2022)

Table 6 Some awareness-raising campaigns developed at Medicus Mundi Alava

Campaigns
"This Christmas ... give social and environmental justice!"[a] FT Christmas campaign carried out by Medicus Mundi Alava every year, from December 1st to January 5th Objective: To disseminate FT and make the store known among the local population Products: Typical Christmas sweets (nougats, panettones, chocolate, etc.) and coffee (100% Arabica), biodegradable and compostable cups made from organic bamboo, and organic cotton and FT pajamas There are handicrafts, textiles, complements, accessories from various countries, cooperative games, books, stationery (calendars, diaries, etc.), organic and natural cosmetics, and gift vouchers
"This Christmas, the baskets carry the FAIR THING" Christmas Basket Catalog, with varied packs composed of quality and FT products with which to surprise and collaborate, improves the living conditions of many people, families, and communities Objective: To promote responsible and fairer consumption
"This Christmas, choose the fair gift!" FT solidarity products ("on the one hand, they are an excellent gift; on the other hand, they carry a lot of personal stories inside"). Each product is quality assured and helps improve many people's lives, families, and communities
"**Multiply X 0**" (Multiply by zero) campaign This campaign aimed to make visible those stores that offered FT products. It has been running for several years The name of the campaign means that FT products, by bringing social and environmental benefits, when consumed, "multiply by zero" inequality, child exploitation, injustice, and unsustainable exploitation of the environment 19 establishments in Vitoria (a Spanish city) joined the campaign. They were identified by a yellow sticker placed on the entrance door, with the campaign slogan This initiative was carried out by the Vitoria Fair Trade Consortium, which, in addition to Medicus Mundi Alava, included other NGDOs (Oxfam Intermon, Bide Bidean, and SETEM Hego Haizea) and was supported by the City Council Objectives: (1) To bring citizens closer to the ethical, environmental, and social values associated with this type of consumption and (2) to demonstrate that in addition to NGDO points of sale, Fair Trade products are present in an increasing number of establishments: food stores, cafeterias, restaurants, etc. To publicize the initiative, press conferences have been held, and it has been disseminated through the networks and websites of the participating organizations. Mini-workshops and tastings have also been held in the stores, as well as basket raffles among customers (to encourage sales in these establishments). A route was also made on Google Maps to locate the stores This campaign has been running for several years, to make people aware of the establishments where they can find FT products
"**World Fair Trade Day**" It celebrates it together with other organizations in the city, in May. It uses the material produced by the State Coordinating Committee of FT, as well as street activities

Source: Medicus Mundi (2022)

[a]Video Example (from the 2017 Christmas campaign): https://www.youtube.com/watch?v=Lm1Q9qNxGZg

In these awareness-raising campaigns, efforts are intensified, and as dissemination channels, Medicus Mundi Alava uses mainly social networks (Facebook, Twitter, Instagram) but also a physical annual newsletter, a blog on its website, other allied websites (e.g., FAMME—Federation of Associations of Medicus Mundi), a Mailchimp newsletter, emails, contests (in the media, such as Radio Vitoria, with questions related to FT), interviews in the media, by Whatsapp, physical supports on the sidewalk at the entrance to the store, press conferences to present the campaigns, TV screen in the store, etc. And as for the materials, it is launched from posters to catalogs (printed—offline and online), gifts, images, photos, and videos for social networks, AdWords, QR codes in the store, video capsule and brochure for companies (considering CSR), a FT mascot for children's audiences, etc.

However, this NGO believes that the way to raise awareness and educate the public is through direct contact. This translates into a large number of face-to-face activities that take place throughout the year, either in the physical store or in spaces that require their presence (e.g., cooking workshops with FT products and tastings of FT products), label reading workshops in the same spaces, dramatized tours of responsible consumption establishments, various contests, webinars on positive experiences of FT at the institutional and company level to encourage others, "escape hall" activities, talks to disseminate the principles of FT, etc.

3 Conclusions

Fair Trade is a global movement that promotes an alternative type of trade to the conventional one and contributes to sustainable development. On the one hand, it is based on sustainable modes of production and marketing, which give priority to people (social benefits) and the planet (environmental benefits) rather than the economy (purely economic benefits); on the other hand, it promotes responsible consumption, raising awareness and sensitizing the population to the importance of their consumption habits in order to achieve social transformation and change the economic model (toward a more sustainable, transparent, supportive, and fair model).

Initially, the FT began with the commercialization of handicrafts, but later, other products, such as food, cosmetics, toys, books, stationery, etc., were included. In addition, within the line of handicrafts, the variety of products has been expanding: textiles (clothing and home), decoration, bijoux/jewelry, etc.

Furthermore, specialized FT stores are no longer the only points of sale for these products: the presence of FT products in conventional stores is increasing, and the proportion has been reversed (now, supermarkets and department stores are the main sales channel for FT products). Online stores are also growing and acquiring greater importance.

Although at first FT could be considered a fad or an interesting commercial opportunity due to the positive attitude that citizens have toward this type of products, FT actually adds value to its products: on the one hand, the quality of the products is taken care of (looking for them to be traditional, healthy, natural, and even ecological products); on the other hand, there are stories and philosophy behind these FT products, which supports vulnerable populations and tries to avoid child exploitation, providing social and environmental benefits, and not only economic benefits.

Medicus Mundi, an NGDO that seeks to promote "Health for All," understands health broadly. Therefore, from a holistic perspective, it seeks to analyze and work on the structural causes that determine and affect people's health. Thus, it has developed a line of work in Fair Trade because it is aware that consumption is a powerful tool for social transformation and that citizens are key to achieving responsible consumption, which helps to eradicate poverty, which ultimately affects people's health.

In short, Fair Trade can be a key instrument for achieving, or at least contributing to, the Sustainable Development Goals (SDGs) (and particularly SDG 12—responsible consumption and production) and sustainability in general.

> *Many small people, in small places, doing small things, can change the world* (Medicus Mundi).

4 Discussion Questions

1. What is your opinion about Fair Trade? Do you think that it is really, in practice, an alternative system to conventional trade?
2. Have you ever consumed any FT products? If yes, explain what your experience has been like/If not, please explain why you have not consumed this type of product.
3. Would you really be willing to pay more for FT products? Why?
4. What do you think the typical consumer of FT products is like?
5. What would you recommend making Medicus Mundi's communication strategy in relation to FT products more effective?
6. How is the FT situation in your country? Do you know of any awareness campaigns on FT?

Acknowledgments We would like to thank the Coordinadora Estatal de Comercio Justo (its English translation would be "State Coordinator of Fair Trade") as well as Medicus Mundi Alava, and especially Marian, for all the information they have provided us with to be able to make this case study.

Appendix: Teaching Notes for the Case

Synopsis

Fair Trade (FT), which emerged in the mid-twentieth century, is a global movement that promotes an alternative type of trade to conventional one: it seeks greater equity in international trade and bases commercial relations on dialogue, transparency, and respect, thus contributing to sustainable development. It is based on production and marketing methods that give priority to people (social benefits) and the planet (environmental benefits) rather than to the economy (maximization of economic benefits). It currently includes the marketing of food products, handicrafts (textiles—clothing and household items—bijoux and jewelry, accessories, decoration), cosmetics, cleaning products, toys, books, and stationery.

This case focuses on Medicus Mundi, a non-governmental development organization (NGDO) that, in order to achieve its goal of "Health for All," aims to eradicate poverty so that everyone can live in dignity and enjoy the right to health in its broadest sense. To this end, in addition to development and humanitarian aid projects, as well as educational initiatives for social transformation, this NGDO fosters Fair Trade. In this way, by promoting quality products (traditional, natural, and, in a significant percentage, organic ones) and an economic model that ensures decent working conditions and remuneration, this organization contributes to sustainability and the achievement of different SDGs (in particular, SDG 12—responsible production and consumption).

Potential Audience and Instructor's Material

This case has been developed for use in the framework of different courses, such as marketing management or nonprofit marketing, or those courses where, in a transversal way, sustainability, responsible consumption, responsible production, or Fair Trade are studied, also in specific seminars on Fair Trade. Thus, the potential audience for this case study is twofold:

- Undergraduate students pursuing degrees in business administration or marketing
- Graduate students pursuing MBA and Masters in marketing, corporate social responsibility, or international cooperation for development; or executive programs and seminars (about marketing in general and Fair Trade in particular).

Learning Objectives

1. To know what Fair Trade is and how it can contribute to responsible production and consumption, and sustainability
2. To study the situation of Fair Trade in Spain

3. To analyze the NGDO Medicus Mundi and, mainly, the activities related to Fair Trade that it develops

Time Frame for Class Discussion

This case can be taught in one session of 60–100 minutes, depending on the number of students participating in the discussion activities as well as the previous work that students have done before class.

Suggested Discussion Questions

1. What is your opinion about Fair Trade? Do you think that it is really, in practice, an alternative system to conventional trade?
2. Have you ever consumed any FT products? If yes, explain what your experience has been like/If not, please explain why you have not consumed this type of product.
3. Would you really be willing to pay more for FT products? Why?
4. What do you think the typical consumer of FT products is like?
5. What would you recommend to make Medicus Mundi's communication strategy in relation to FT products more effective?
6. How is the FT situation in your country? Do you know of any awareness campaigns on FT?

Analysis

Answer to Question 1. *What is your opinion about Fair Trade? Do you think that it is really, in practice, an alternative system to conventional trade?*

These are open-ended questions for students to give their opinions on this topic, reflecting on the pros and cons, and comparing both types of trade (Fair Trade vs. conventional trade).

Answer to Question 2. *Have you ever consumed any FT products? If yes, explain your experience has been like/If not, please explain why you have not consumed this type of product.*

This is an open-ended question for students to explain their experiences related to FT products and the reasons why they do, or do not, consume FT products.

Answer to Question 3. *Would you really be willing to pay more for FT products? Why?*

This is also an open-ended question for students to reflect on whether they would be willing to pay more for FT products (their retail value may be higher, as they guarantee fairer remuneration and better working conditions for producers).

It is often easy to say "yes" in surveys, but then this behavior does not occur (e.g., if consumers are price sensitive or do not have sufficient purchasing power: although they find FT products interesting because of the values they defend, in the end the price weighs more heavily in their decision, and they buy those products that are more affordable for their pocket).

Answer to Question 4. *What do you think the typical consumer of FT products is like?*

This open-ended question is for students to reflect and investigate what profile or profiles FT consumers have.

They could look in search engines such as Google or in academic publication databases, for papers focused on the FT consumer.

Answer to Question 5. *What would you recommend making Medicus Mundi's communication strategy in relation to FT products more effective?*

This open-ended question seeks to get students to work from the perspective of the NGDO. It also seeks to get them to think about the importance of communication strategy in relation to FT products and how its design can affect its effectiveness.

Students can also propose different communication strategies and discuss the pros and cons of each.

As an example, students can watch the video at https://www.youtube.com/watch?v=Im9J0uqjIHw or another similar video.

Or this: https://vimeo.com/192599128

Answer to Question 6. *How is the FT situation in your country? Do you know of any awareness campaigns on FT?*

This is an open-ended question. The purpose of this question is for students to investigate the situation of Fair Trade in their country and to make comparisons with the case described (Fair Trade in Spain): whether there is more or less knowledge of Fair Trade in their country, the volume of sales of these products (more or less than in Spain), the awareness campaigns that are carried out, etc.

References

CECJ – Coordinadora Estatal de Comercio Justo. (2022). Official website. Available from https://comerciojusto.org/

Fairtrade International. (2022). Official website. Available from https://www.fairtrade.net/

Fernandez-Ferrin, P., Galan-Ladero, M. M., Calvo-Turrientes, A., & Castro-Gonzalez, S. (2021). *El Comercio Justo en Euskadi. Comportamientos y Motivaciones de Compra*. Vitoria (Spain): Medicusmundi Alava/Araba. Available from https://comerciojusto.org/publicacion/el-comercio-justo-en-euskadi-comportamiento-y-motivaciones-de-compra/ https://comerciojusto.org/wp-content/uploads/2022/03/investigacion-cj-en-euskadi-def.pdf

Medicus Mundi. (2022). Official website. Available from https://www.medicusmundi.es

Tapia-Mantilla, R. (2012). Economía solidaria: una alternativa de Desarrollo integral en América Latina. *El Portal de la Economía Solidaria*. Available from https://www.economiasolidaria.org/noticias/economia-solidaria-una-alternativa-de-desarrollo-local-integral-en-america-latina/

WFTO – World Fair Trade Organization. (2022). Official website. Available from https://wfto.com/

The manufacturer's authorised representative in the EU is Springer Nature Customer Service Centre GmbH, Europaplatz 3, 69115 Heidelberg, Germany. If you have any concerns regarding our products, please contact ProductSafety@springernature.com

Printed and bound by CPI Group (UK) Ltd, Croydon, CR0 4YY
01/10/2026
02225872-0007